imagine darkness

Also by Charles Scurlock

the picnic at the edge of the universe
2011, 2015

imagine darkness

the making of *the simple universe*

charles scurlock

imagine darkness

© Copyright 2015   Charles Scurlock

Charles Scurlock
imagine darkness
the making of the simple universe

www.enquiriesnw.com

ISBN 978-1508479567

Acknowledgements and author's note:

This book is an extension and, in parts, a completion of the work I sketched out in *the picnic at the edge of the universe*, published as an ebook in 2011. Those few of you who took a chance that I was able to write something readable and read that book will notice a few of the same references and quotations. They are here because they are important to me and a part of why both books came into existence. In particular, those are the brief quotes from Janna Levin, John R Platt, and Elizabeth Rosner, words that helped me crystallize my thinking at an important moment, so I've taken the liberty of repeating them here.

As for more personal acknowledgements, I can do no better than to repeat the list I gave in that first book:

Thanks–

–to Larry Durocher, Herb Helsel, Billy Kerby, Warner Scheyer, Doug Struthers and Doug York (the other Doug!) for reading, commenting, discussion, and debate;

–and to Ursula, my truly good wife and best supporter, editor, and critic.

You have all been extraordinarily patient and tolerant of my rants and jibes. Thanks again.

I also want to give particular thanks to the makers and supporters of the internet, and Wikipedia, and all those who make a point to post books, and archives, and scientific (even some not so scientific) papers there for all of us to access, for without them I would be spending several more years digging through old musty records and libraries for what I have been able to track down in just these two or three.

—cs, January 2015

In my dream, the angel shrugged, and said, 'If we fail this time, it will be a failure of imagination;'
—and then she placed the world, gently, in the palm of my hand.
  —Brian Andreas

Contents

Preface

THE ORIGINS OF THIS BOOK ARE TWO. The first goes back a very long way, probably to high school physics, my introduction to the laws of thermodynamics. The long, slow slide into entropy. The heat death of the universe. Absolute zero. Why, I wondered, if this were true, if all of this had been here for so many billions of years, how could it be that there was still something rather than nothing?

The second and more proximate cause was a question asked in Janna Levin's 2003 book, *How the Universe Got Its Spots*, (1) Levin, an astrophysicist at Columbia University, posed it very simply. "Is the universe infinite, or is it just really, really big?" A simple question but with huge implications.

Infinite I rejected out of hand. It's just a word to denote unknown, unmeasurable. It has meaning only in mathematics, not in reality. If the universe is not infinite, then out there somewhere there is an edge, a boundary. We're told it's expanding. How do we know that, and if it is, into what? And if it's expanding then it must have started up sometime in the past. Where, when, how? I've always liked questions more than answers, so that's where it all began. It led down many paths, first resulting in my first book, that now seems only a sketchy outline, *the picnic at the edge of the universe*, in 2011, and then into more years of reading and research, to determine if those ideas could really lead me into something substantive. This is the current outcome of that effort. It is not yet and perhaps may never be complete, but that's the way of science, and of physics and cosmology in particular.

* * * * *

Thomas S. Kuhn, in *The Structure of Scientific Revolutions*, (2) suggests that most such revolutions result from the appearance of serious anomalies between observations and the prevailing paradigm or reigning theory of a particular discipline. As examples in physics, he cites the Copernican revolution, which overturned the Ptolemaic model of the universe which had held sway for over 1500 years. In this case the anomalies were observations that while for the most part the wandering stars, our planets, behaved in general accordance with the platonic ideal of perfect circles and consistent behavior, some appeared to misbehave by moving backwards for a while before returning to their normal paths. Ptolemy had responded to this by adding little fixes, orbital epicycles to fit observations. The fact that the other observed orbital anomalies did not precipitate a crisis sooner had other causes as well, the power of the church being a major one, with its insistence that man, and hence the earth, had been made the center of the universe by the hand of God. Dispute that at very high cost, they said. But overturn it Copernicus did, thereby launching a major restructuring of scientific thought.

Isaac Newton is the second of Kuhn's examples, when his mechanics and mathematics replaced the Aristotelian paradigm that had also stood almost inviolate for all those centuries. Prime among Newton's advances was the identification of gravity as the force that caused objects to fall to earth, not their "affinity" for earth, a concept notable for vagueness, but in line with most thinking of the time.

Newtonian physics was enough for a long time, until the growth of astronomy began to overtake his concept of gravity as it might apply to those distant objects we began to observe. Einstein took care of that with a double barreled blast in 1905 and 1916 with first Special Relativity and then General Relativity, replacing Newton's fixed ideas of space and time with what he called a four-dimensional continuum that had real physical qualities and could be curved, stretched, and compressed. Again, according to Kuhn, the observed anomalies had accumulated to a point of crisis, and the world of physics was ready for a new paradigm.

A funny thing happened here, however. Another revolution was in progress. Einstein had apparently taken care of the universe, the sun, the stars, the galaxies, and how they all held together, but at the other end of the scale, where a new chemistry and a new physics were arising in the form of the so-called "quantum" theories, relativity did not have answers. The curvature of spacetime was insufficient to explain how the smallest part of the universe was held together. Einstein was fully aware of this, and he along with many others worked for many years to reconcile the existence of what they described as the four fundamental forces of physics, the strong force, electrodynamics, the weak force, and the weakest of all, gravity. And that is where the revolution in physics stands today, with a still strong belief that there must be a way to resolve these discrepancies, but no clear way to the goal.

A key part of Kuhn's argument is that each of these revolutions resulted in a change in scientists' conceptual visions, a different way of seeing old familiar data and observations, seeing with new eyes, or as if they were visitors to this planet and had never encountered these experiences before.

In the instance of the Copernican revolution, it was simple but profound, the realization that we were seeing the universe from a moving platform, not a stationary one. This idea had many opponents because it challenged over a thousand years of common sense, and required the rewriting of most accounts of creation.

Newton also changed how we see the world. Here on earth we had to abandon what I call the granting of emotions to material objects. They no longer had an *affinity* for the earth, a desire to be reunited with it, drawing them down. Rather, both they and the earth were in the grip of a force that moved them together. He showed us ways to measure that force. He also showed us that the orbits of the planets were a result of that same force, drawing them into circumferential paths balancing the centrifugal force that would otherwise have carried them off into outer space.

Einstein changed our conceptual vision into one even more abstract, by making mathematics the heart of it rather than just a description to explain it. His Special Relativity formalized concepts earlier explained to us by Galileo and others, that smooth, even motion was indistinguishable from no motion, absent outside, observable reference frames. He made it possible for us to imagine that we are actually moving at about a thousand miles per hour on this spinning globe even though we can't feel it. But perhaps more importantly, he gave us $E=mc^2$, a simple equation that quantified the equivalence of matter and energy and led to nearly every new concept in physics for the next hundred years. Suddenly science saw that part of the world with new eyes. Then, about ten years later, his Theory of General Relativity, consolidating the work of many other theorists, changed how we imagined space and time.

This story has been told and is still being told over the last hundred years in literally thousands of books, papers by eminent experts, movies, now even web sites, and most recently by the promulgation of dozens of new speculative theories. Unfortunately, the majority of these fall into the category that one critic has dubbed "fairy tale physics," since they share a particular characteristic, they are based mostly on unproved and unprovable assumptions, and are not supported by any reliable research or observations. One set of these, the so-called "string theories," has been around for more than thirty years but has as yet produced no testable assertions. It has been described by one eminent critic as being based on something like "throwing a dart at a wall and then drawing a target around the point of impact."

Unfortunately, where we find ourselves today is with two currently accepted narratives that are separately known as "Standard Models," one of physics, and one of cosmology. Let's talk a bit about them.

The first is the *Standard Model of Particle Physics*, the generally accepted set of theories that Wikipedia, the open source encyclopedia of human knowledge on the world-wide web describes this way:

"The **Standard Model** of particle physics is a theory concerning the electromagnetic, weak, and strong nuclear interactions, which mediate the dynamics of the known subatomic particles. It was developed throughout the latter half of the 20th century, as a collaborative effort of scientists around the world. The current formulation was finalized in the mid-1970s upon experimental confirmation of the existence of quarks. Since then, discoveries of the top quark (1995), the tau neutrino (2000), and more recently the Higgs boson (2013), have given further credence to the Standard Model. Because of its success in explaining a wide variety of experimental results, the Standard Model is sometimes regarded as a "theory of almost everything"."(3)

This definition makes reference to a number of entities about which the average reader may actually know very little. You may ask, "What's subatomic particle? Aren't atoms supposed to be the smallest things of all? What's a neutrino, a quark, a Higgs boson?" Those are excellent questions and we'll try to explain some of them, not all, and we'll refer you to other sources that will help, as far as they can go.

Wikipedia goes on, of course, because in spite of TSM's reputed success, noted in the last sentence above, there are some minor and some major questions about it:

The Standard Model falls short of being a complete theory of fundamental interactions because it makes certain simplifying assumptions. It does not incorporate the full theory of gravitation as described by general relativity, or predict the accelerating expansion of the universe (as possibly described by dark energy). The theory does not contain any viable dark matter particle that possesses all of the required properties deduced from observational cosmology. It also does not correctly account for neutrino oscillations (and their non-zero masses). Although the Standard Model is believed to be theoretically self-consistent and has demonstrated huge and continued successes in providing experimental predictions, it does leave some phenomena unexplained.

The development of the Standard Model was driven by theoretical and experimental particle physicists alike. For theorists, the Standard Model is a paradigm of a quantum field theory, which exhibits a wide range of physics including spontaneous symmetry breaking, anomalies, non-perturbative behavior, etc. It is used as a basis for building more exotic models that incorporate hypothetical particles, extra dimensions, and elaborate symmetries (such as supersymmetry) in an attempt to explain experimental results at variance with the Standard Model, such as the existence of dark matter and neutrino oscillations." (4)

Ptolemy lives! Epicycles are everywhere!

Thomas Kuhn describes an accepted paradigm in science as a more or less complete theoretical model that provides a more or less stable platform for what he calls normal science By this he means the work in nature and the laboratory that engages the worker bees of science in the job of confirming theoretical concepts, measuring and testing the mathematical basis of the theory against the reality of nature. And finally, drawing conclusions that inventors and engineers can reliably depend on for real world applications. This of course is what has occurred historically with every revolution we referred to earlier. Through that process, which may last for many years, anomalies will also arise, perhaps not to the level of crisis. Many of those will be resolved by minor adjustments to the theory. We lived with Ptolemy's epicycles for centuries after all, didn't we? But some will stay unresolved. In quantum theory for example, some anomalies have been carried along as what many would call paradoxes of the theory, some as obvious logical contradictions, some as almost mystical explanations of phenomena that cannot be observed but only presumed to occur. Examples, *wavefunction collapse*, "virtual" particles and the like, about which we will talk later. A question you will also find asked later is "Who has seen a hadron?" The Standard model claims to have identified at least 61 fundamental particles that are assumed to make up all matter in the universe. This number does not include the full set of antiparticles that are required to make the mathematics of many observations work correctly. Then there are the *virtual* ones that are so named because the math needs them but they seem to stay around for too brief a time to be observed. As you may begin to see, there are too many questions to be resolved in this brief introduction.

The second Standard Model is the one that supposedly describes the Universe as a whole along with its origins, as opposed to its smallest parts, the domain of Standard Model 1. This is the one that is best known by its popular title, "The Big Bang." This model also includes most of Einstein's two theories of relativity. Wikipedia gives it this introduction:

"The **Big Bang** theory is the prevailing cosmological model for the early development of the universe. The key idea is that the universe is expanding. Consequently, the universe was denser and hotter in the past. In particular, the Big Bang model suggests that at some moment all matter in the universe was contained in a single point, which is considered the beginning of the universe. Modern measurements place this moment at approximately 13.82 billion years ago, which is thus considered the age of the universe. After the initial expansion, the universe cooled sufficiently to allow the formation of subatomic particles, including protons, neutrons, and electrons. Though simple atomic nuclei formed within the first three minutes after the Big Bang, thousands of years passed before the first electrically neutral atoms formed. The majority of atoms that were produced by the Big Bang are hydrogen, along with helium and traces of lithium. Giant clouds of these primordial elements later coalesced through gravity to form stars and galaxies, and the heavier elements were synthesized either within stars or during supernovae.

The Big Bang theory offers a comprehensive explanation for a broad range of observed phenomena, including the abundance of light elements, the cosmic microwave background, large scale structure, and the Hubble diagram. The core ideas of the Big Bang—the expansion, the early hot state, the formation of light elements, and the formation of galaxies—are derived from these and other observations. As the distance between galaxies increases today, in the past galaxies were closer together. The consequence of this is that the characteristics of the universe can be calculated in detail back in time to extreme densities and temperatures, while large particle accelerators replicate such conditions, resulting in confirmation and refinement of the details of the Big Bang model. On the other hand, these accelerators can only probe so far into high energy regimes, and astronomers are prevented from seeing the absolute earliest moments in the universe by various cosmological horizons. The earliest instant of the Big Bang expansion is still an area of open investigation. The Big Bang theory does not provide any explanation for the initial conditions of the universe; rather, it describes and explains the general evolution of the universe going forward from that point on." *5)

So, a few more terms that will need more explanation, which is forthcoming but not needed here. But it gives you a sense of the complexity of the subject as it is now conceived. I'd like to add at least one more layer of detail here so that we can come back to this later in our discussion of alternative beginnings for comparison.

"The model includes a single originating event, the "Big Bang" or initial singularity, which was not an explosion but the abrupt appearance of expanding space-time containing radiation at temperatures of around 10^{15} K. This was immediately (within 10^{-29} seconds) followed by an exponential expansion of space by a scale multiplier of 10^{27} or more, known as cosmic inflation. The early universe remained hot (above 10,000° K) for several hundred thousand years, a state that is detectable as a residual cosmic microwave background, or CMB, a very low energy radiation emanating from all parts of the sky. The "Big Bang" scenario, with cosmic inflation and standard particle physics, is the only current cosmological model consistent with the observed continuing expansion of space, the observed distribution of lighter elements in the universe (hydrogen, helium, and lithium), and the spatial texture of minute irregularities (anisotropies) in the CMB radiation. Cosmic inflation also addresses the "horizon problem" in the CMB; indeed, it seems likely that the universe is larger than the observable particle horizon." (6)

The two key aspects of the big bang are one, the mother of all explosions that initiated it, demurrals by the Wikipedia author notwithstanding, and the Einsteinian, and two, the relativistic nature of how it evolved and which describes the relationships between and among its evolved constituent parts. As we have noted before, *nothing* can be known before a beginning, by definition, but with an exception. If something might have existed before a particular event and evidence of it still exists, then that beginning must have been preceded by something. If our universe is finite, and exists and is expanding into that something, call it space, or a field, or whatever, then a new look is in order.

The big bang idea was first proposed by Georges Lemaitre, in 1927(7), though he didn't give it that name. The concept built over time into the broadly based construct we know today. The framework for the Big Bang model relies on Albert Einstein's general relativity and on simplifying assumptions such as the homogeneity and isotropy of space, the "cosmological principle" that space has the same characteristics throughout, that it looks the same no matter from what point we might view it.

The "Big Bang" model has not been unchallenged. Even its name came not from a positive reference but from an author of what was then a strong competitor, the "Steady State" cosmology, proposed by the astronomer Fred Hoyle and two co-authors. Hoyle, in an interview recorded in the heat of the debate over the two theories derisively called his competition a "big bang, " and the name stuck.

Of course almost all theorizing about what we cannot know directly depends on simplifying assumptions. However, in this case we have gone all the way to the most difficult off assertions to accept, contradicting every scientific authority back to and including the earliest recorded history, always excepting divine intervention, that the universe was generated out of nothing. Confirmations of the big bang have come from respected astronomers and refutations of it have come from respected astronomers. Many observations appear to confirm it as a plausible explanation while others are not explained. One astronomer, T. von Flandern, has published *The Top 30 Problems with the Big Bang (8)*, listing those either not explained or apparently not explainable by more plausible causes or events. But the theory's acceptance has been sufficiently general that opposing theories have found little attraction among academic researchers.

And, of course, if it is based strongly in Einstein's relativity, the simplifying assumptions of those theories must also come under scrutiny. Many scientists did so early on, when those were first published, and many still question them today. You will find them questioned here, particularly in their objectification, primarily in giving real physical attributes to two purely theoretical concepts, those of *space* and *time*, not to speak of the slowing of time and the shortening of rulers at near light speeds.

As you read further, you will find us questioning these and many other aspects of today's commonly accepted models of the universe. For the most part, they will be challenged not on the basis that they are intentionally false, but that they are based on either misinterpreted or misattributed observations and data. Some will be challenged for the common assumptions, particularly by mathematical physicists, that mathematics is not just a describer but actually is the reality. Some will be challenged because they are actually more mystical than objective in nature, a position that has no place in our preeminently most *physical* of the sciences. Another way of saying this is to ask science the question: " When we look at what we perceive as the universe, what are we actually seeing?" Is the way we have looked at it up until now the only way to explain what we see? Is there not another way to look?

This book is not just a challenge to the accepted theories. It is intent is to offer a clear alternative model, *the simple universe*, a model that explains the observations with a paradox-free, contradiction-free, and logical structure. The goal is to describe a universe free of unnecessary complications. A simple universe understandable to all. A model that stands on its own without mystical underpinnings requiring giant leaps of faith or *willing suspensions of disbelief*, in the words of Coleridge. Will it be complete? As close as we can make it in one short volume. Will it answer all the questions? No. I am far from that level of knowledge or arrogance. Will it start a scientific revolution? One can only hope. But it will, I am convinced, provide a place to start, a new way of seeing and thinking about the facts, most of which we already know but have been looking at as if we were metaphorically still on that unmoving, stationary earth of Ptolemy's time. Let us embark.

We started by repeating Janna Levin's question about the possible infinitude of the universe and the choice to go with finite instead. In her second book, *The Mad Man Dreams of Turing Machines*, she offers another way to look at the facts:

"There are faint stars in the night sky that you can see but only if you look to the side of where they shine. They burn too weakly or are too far to be seen directly, even if you stare. But you can see them out of the corner of your eye because the cells on the periphery of your retina are more sensitive to light. Maybe truth is just like that. You can see it, but only out of the corner of your eye."[1(9)]

Part 1. The Parts—The Universe as an Ill-structured Problem

"The central task of a natural science is to make the wonderful commonplace: to show that complexity, correctly viewed, is only a mask for simplicity; to find pattern hidden in apparent chaos."
—Herbert A. Simon

imagine darkness

1.1 Introduction

IMAGINE DARKNESS. SILKEN. DEEP. INTENSE. A dark so dark you almost feel as well as see it. The dark you see right after you put down your book, switch off the bed lamp, close your eyes. A dark marked only by the flickering lights of the few neurons still firing in the visual cortex before your brain relaxes into sleep. Imagine, if you can, a *shimmering* darkness, its motion well below the threshold of your consciousness, but still you know it's there.

Imagine silence. The silence of a moment in an anechoic chamber, those rooms used by acousticians to test new devices, a place where uttered sound goes out but nothing returns. Imagine just a faint distant noise like static, detectable only by the most sensitive instruments.

Imagine emptiness. Nothing to touch, taste, feel. Imagine no such thing as time or space, since nothing is there to happen or inhabit, nothing to be observed or measured or described.

You are in an ocean. An ocean without shorelines, without top or bottom. Not *infinite,* that mathematical concept of the unmeasurable, just very, very big. An ocean of energy, of high frequency, high entropy, and just those flashes, those tiny evidences of turbulence, currents, eddies.

* * * * *

There is an "in the beginning" in every creation myth, from the King James Bible, to the Hindu Vedas, to the Big Bang.

This is where *we* start, this darkness, silence, emptiness. Can everything we know in the world have come out of this? Can we build from this beginning a model of the universe, maybe more than one, from just this thin, evanescent soup of energy that fills what has been called space, the void? Can we build from what we know of how things work, from the insights of men and women like Newton, Einstein, and other geniuses who preceded us as well as from what we have observed since their time? Can we use what of theirs has been shown to be true, and can we see past their errors and unknowns to draw a new, consistent, and more complete model of the world?

I think we can. What it will take is to step back and see the world as if it were new, to see fresh the observations we have misunderstood, misattributed, misconstrued, and put them into a new model without the paradoxes, contradictions, and myths that characterize the old ones. A new model wholly based in reality, not based, as one recent writer has described the standard models of physics and cosmology, on unsupported speculation on unsupportable assumptions on prior unsupportable assumptions.

Everything we have to say in the work following, from the birth and growth of the smallest entities of our universe to the origin and development of its largest and most complex of them, the stars, galaxies and clusters we see at such great distances, depends on the existence of a primal lightless, soundless, (almost) empty universal electromagnetic field. That is the only thing must for now take as given, as pre-existing everything else. And while there is as yet nothing we can say about its origins, how it came to exist, we have ample evidence that however it came about, that it does exist, that it is here, there, and everywhere, pervading everything that has arisen out of it and the perceived emptiness that surrounds them. So all of the mysteries, paradoxes, and contradictions of modern physics and cosmology, from quantum theories; multiverses, dozens, even hundreds of dimensions, "singularities," can, in the end, be reduced to just this one.

What can we know about this field? Well, we can observe it, here, in the zone of middle dimensions, as what happens around a magnet, what carries our radio, wireless, television dramas and data and information. We can feel its energy on a sunny day, we can see the changes it makes in the color of our skin. We can see how its energy causes the growth of plants. We use it to heat our houses , drive our vehicles, and feed our bodies. And from that knowledge we can see it as an electromagnetic field, something we actually know something about.

From what we know of the character and velocity of light and how it is promulgated, we can infer the probable native frequency of the field, likely to be in the neighborhood of $1/h$. From astronomical observations of the bending of light around high energy entities and from the discovery of what astronomical mystery lovers have called "dark matter" we can infer the presence of distortions of the field and those distortions' effects on other entities in it. From the detection of the electromagnetic background noise of what has been called cosmic microwave background radiation we can determine the average energy level of the field, its turbulent nature, and its pervasiveness in and throughout the universe and its surroundings. We can study the distortions of the field and derive clues toward the simplification of the forces and interactions that have so far eluded all of physics' efforts to parse and understand, from so-called subatomic forces up to and including magnetism and gravity. We can begin to see patterns and derive from them clues to a new set of laws and rules that govern all physical processes from the smallest to the largest scales, much as we have seen in the commonality of scale demonstrated in fractal geometry. And we will be able to abandon the quasi-mystical assertions of the quantum theorists, like "wave-particle duality" and "wavefunction collapse." And Schrödinger's cat will no longer find it necessary to maintain the illusion that it is simultaneously both alive and dead.

In the words of John R. Platt, from the journal *Science*, in 1964:

> *"Many—perhaps most—of the great issues of science are qualitative, not quantitative, even in physics and chemistry. Equations and measurements are useful when and only when they are related to proof; but proof or disproof comes first and is in fact strongest when it is absolutely convincing without any quantitative measurement.*
>
> *—to say it another way, you can catch phenomena in a logical box or in a mathematical box. The logical box is coarse but strong. The mathematical box is fine-grained but flimsy. The mathematical box is a beautiful way of wrapping up a problem, but it will not hold the phenomena unless they have been caught*

in a logical box to begin with."

I am not a mathematician, so you will not find here a mathematical box. We leave that to others. The goal here is simply to construct a strong logical box, perhaps one sturdy enough to one day serve as a suitable container for a (dreamed of) theory of everything.

imagine darkness

1.2 Rediscovering the Ether

"In the beginning God created the heaven and the earth.

And the earth was without form and void, and darkness was on the face of the deep. And the Spirit of God moved upon the face of the waters.

And God said, Let there be light: and there was light."

—*Christian Bible, King James version, Genesis, Chapter 1, verses 1-3*

In every creation myth there is a beginning. And because it is a beginning, we will most likely never know of what came before, because, by definition, there was nothing before. So no matter what our conception, no matter what our model of the universe, there is no real knowledge of a true origin. We are, of necessity, perhaps, left with speculation. Unless, of course, that what was there before is still with us. In our currently accepted creation myth, the so-called Big Bang theory, we have struggled to imagine a beginning that might explain what we can see in the world and outside of it, both with our human senses and our instruments that extend those senses out into the distances of space and the recesses of time. That struggle has led to a mostly quiet collaboration between the disciplines of physics, astronomy, and cosmology; physics, charged with finding explanations of how the world is made and how it works; astronomy, with its attempts to see, record, and understand the workings of those distant bodies and systems that surround our planet; and cosmology, responsible for linking those efforts and, hopefully, delineating theories that might unify our understanding of the local and the distant phenomena.

We find ourselves today, however, in a dilemma. In the world of theoretical physics, we have arrived at a so-called standard model that, while it seems to work in a more or less practical sense, is riddled with logical paradoxes and contradictions, and as a result, must be considered at best incomplete and at worst as being totally wrong in its conception and application. In the world of astronomy , though blessed with the most powerful and capable instruments and methods ever known, we find ourselves still searching for a model that comes close to explaining often disconnected and/or contradictory observations, among them accurate distance estimates, expansion or not, behavior of galaxies, clusters, quasars, and the like. In cosmology, while the Big Bang seems generally accepted as the "beginning" model of choice, it faces constant criticism as requiring huge logical leaps of faith as to its cause and origin. How could the universe start from nothing? What is this thing called a "quantum fluctuation"? And if it actually happened some 13.8 billion years ago, why are there so many discontinuities in what we observe in the skies, like galaxies that appear to be much older?

One result of this state of affairs is the proliferation of speculative models that seem to have little or no scientific basis, that seem to be just put out there to generate even more speculation.

In the almost one hundred years since Albert Einstein published his General Theory of Relativity, we have seen enormous strides in the practical uses of those ideas and in those of other advances in physics, but there are still great and seemingly insurmountable gaps between the theories of the very large and the very small. Many in the field feel that it is time to take another look at our assumptions, our theories, even our observations and their attributions.

So we are here and now ready to take a step back and see where we might have gone wrong. It seems appropriate to begin at the beginning. If we start by rejecting the idea that anything, particularly something as big and important as our universe could not have come from nothing, then what could it have started from? What could have supplied the enormous amount of energy required to generate the vast and complex organism that we can see and feel around us? It must be something very large in extent and very pervasive and powerful. What can we imagine with that power? Has anyone before us imagined such a thing, outside of some mythical omniscient and omnipotent deity?

If we look back in history there was something that was considered at least as pervasive, if not as powerful, Newton's (and others') Luminiferous Aether, the medium that they thought enabled a vacuum to carry light.

In his long career as an intellectual giant of his time in philosophy and science (particularly in chemistry) Robert Boyle (1627-1691) was the author of numerous tracts, papers and ideas. He is best known for Boyle's law, which describes the inversely proportional relationship between the absolute pressure and volume of a gas, if the temperature is kept constant within a closed system. For our purposes, we are particularly interested in his idea of the aether, which to his mind, in the years before Isaac Newton's adoption of a similar idea, was a *probable hypothesis* and consisted of what he called subtle particles, one sort of which explained the absence of vacuum and the mechanical interactions between bodies, and the other sort explained phenomena such as magnetism (and possibly gravity) that were inexplicable on the basis of the purely mechanical interactions of macroscopic bodies:

"...though in the ether of the ancients there was nothing taken notice of but a diffused and very subtle substance; yet we are at present content to allow that there is always in the air a swarm of steams moving in a determinate course between the north pole and the south." (W) (1)

When Newton began his studies of light, he saw it as being made up of tiny units he called corpuscles, by which he could explain how light could travel in straight lines and be reflected off surfaces, but he struggled with the explanation of the refraction of light through transparent materials. He found the concept of Boyle's ether a useful one to explain the characteristics of particular qualities of light, especially that of refraction. In order to explain refraction, Newton's *Opticks* (1704) postulated an "Aethereal Medium" transmitting vibrations *faster* than light, by which light, when overtaken, is put into "Fits of easy Reflexion and easy Transmission", which caused refraction and diffraction. Newton believed that these vibrations were related to heat radiation.

He wrote:

"Is not the Heat of the warm Room convey'd through the vacuum by the Vibrations of a much subtiler Medium than Air, which after the Air was drawn out remained in the Vacuum? And is not this Medium the same with that Medium by which Light is refracted and reflected, and by whose Vibrations Light communicates Heat to Bodies, and is put into Fits of easy Reflexion and easy Transmission?"

[...] **"I do not know what this Aether is"**, but that if it consists of particles then they must be **"exceedingly smaller than those of Air, or even than those of Light:** The exceeding smallness of its Particles may contribute to the greatness of the force by which those Particles may recede from one another, and thereby make that Medium exceedingly more rare and elastic than Air, and by consequence exceedingly less able to resist the motions of Projectiles, and exceedingly more able to press upon gross Bodies, by endeavoring to expand itself." (2)

Christiaan Huygens, before Newton, saw light as a *wave* being propagated through an ether, a concept rejected by Newton, a preview of a conflict that remains common in physics today. In 1720, James Bradley carried out experiments that appeared to support Newton's corpuscular notion of light. A century later, however, Young and Fresnel revived the wave theory of light and in the wake of a series of experiments on diffraction the particle model of Newton was finally abandoned. Physicists assumed, moreover, that like mechanical waves, light waves required a medium for propagation, and thus required Huygens's idea of an aether "gas" permeating all space. However, a wave apparently required the propagating medium to behave as a solid, that is, non-moving, as opposed to a gas or fluid.

This was the "state of the ether" until the nineteenth century, when first Hertz, then Faraday, then Maxwell developed their theories of electromagnetic fields and phenomena, in which light and heat energy were finally included. Maxwell's work culminated in a set of eight equations, later reduced to four by Oliver Heaviside, based on the work of Gauss, Ampere, Weber, and Kohlrausch. The resulting wave equations represented an electromagnetic wave that propagates at the speed of light, hence supporting the view that light is a form of electromagnetic radiation.

The understanding of electromagnetic waves at this time in history required that they be the result of moving "charged bodies" and hence could not be propagated in a true vacuum. Maxwell's equations also required that electromagnetic waves were propagated at a fixed speed, the speed of light, "c". This brought the whole system, of necessity, into the realm of Galilean-Newtonian relativity, since the fixed speed of light could only be recognized in a 'fixed' reference frame. As this can only occur in one reference frame, in Newtonian physics the aether is hypothesized as an absolute and unique frame of reference in which Maxwell's equations hold. That is, the aether must be "still" universally, otherwise c would vary along with any variations that might occur in its supportive medium. This was understood clearly by Hendrik Lorentz in 1892. (3)

But while the existence of the ether was still not experimentally demonstrated, its concept was adopted (and misused) by numerous non-scientific groups and, although it seemed generally accepted as a given, it was basically ignored in many quarters. Then, In 1887 , Albert Michelson and Edward Morley at what is now Case Western Reserve University in Cleveland, Ohio, carried out what has become famous as the Michelson-Morley experiment. (4) It attempted to detect the relative motion of matter through the stationary luminiferous aether ("aether wind"). The negative results are generally considered to be the first strong evidence against the then prevalent aether theory, and initiated a line of research that eventually led to special relativity, *in which the stationary aether concept has no role.* The perception of the physicists of the day, their "world view", in fact, was that every substance must be made up of "particles" and that light, like every other observable phenomenon, must travel *through* a medium, as a ship travels *through* the ocean, and that if there appeared to be nothing to resist the movement of light, nothing to slow down its passage, then there must simply be *nothing* there! The ether must not exist!

The ether was kept alive, however, although in a new form, by Hendrik Lorentz, who in his Nobel acceptance address in 1902 described it thusly:

"[...] in its annual journey round the sun the earth travels through space at a speed more than a thousand times greater than that of an express train. We might expect that in these circumstances there would be an end to the immobility of the ether; the earth would push it away in front of itself, and the ether would flow to the rear of the planet, either along its surface or at a certain distance from it, so as to occupy the space which the earth has just vacated. Astronomical observation of the positions of the heavenly bodies gives a sharp means of determining whether this is in fact the case; movements of the ether would assuredly influence the course of the beams of light in some way. [...] Thanks to the investigations of Van der Waals and other physicists, we know fairly accurately how great a part of the space occupied by a body is in fact filled by its molecules. In fairly dense substances this fraction is so large that we have difficulty in imagining the earth to be of such loose molecular structure that the ether can flow almost completely freely through the spaces between the molecules. Rather are we constrained to take the view that each individual molecule is permeable. The simplest thing is to suggest further that the same is true of each atom, and this leads us to the idea that an atom is in the last resort some sort of local modification of the omnipresent ether, a modification which can shift from place to place without the medium itself altering its position. Having reached this point, we can consider the ether as a substance of a completely distinctive nature, completely different from all ponderable matter. With regard to its inner constitution, in the present state of our knowledge it is very difficult for us to give an adequate picture of it." (5)

The title of Lorentz' lecture was "The Theory of Electrons and the Propagation of Light." His world view expressed in it was made up of three elements, Ether, Electrons, and Ponderable Matter, so it was constrained by the knowledge available at that time. But his ether was not what Michelson and Morley disproved the existence of. Lorentz' ether was of a finer construction than one of "particulate matter" that would have been displaced by the movement of ponderable matter in that experiment.

So, we come to the next revolution in physics. In 1900, Max Planck published work that resulted in a profound change in modern physics. Planck was studying "black body" radiation, a measurable way of showing a wide range of energy values based on their "color temperatures" as their energy level increased. His struggles and barriers were, however, with the mathematical formulae with which his findings could be expressed. There seemed to be no smooth way to write those equations. Finally, he reluctantly came to a conclusion.

"The central assumption behind his new derivation, presented to the DPG on 14 December 1900, was the supposition, now known as the Planck postulate, that electromagnetic energy could be emitted only in quantized form, in other words, the energy could only be a multiple of an elementary unit $E = hv$, where h is Planck's constant, also known as Planck's action quantum." (Wikipedia) (6)

Newton's "corpuscles" were back, though wearing new clothes. In 1905, Albert Einstein followed with his Special Theory of Relativity which contained, for our purposes, two significant points.. One was his acceptance of Planck's "quantum" nature of the energy of light, (later to be called "photons"). The second was the famous equation $E=mc^2$, which linked forever the relationship between matter and energy, a concept later confirmed by nuclear physicists, and one with profound implications for our model of "the simple universe." As Einstein himself later expressed it:

"A further consequence of the (special) theory of relativity is the connection between mass and energy. Mass is energy and energy has mass. The two conservation laws of mass and energy are combined by the relativity theory into one, the conservation law of mass-energy."
"From the relativity theory, we know that matter represents vast stores of energy and that energy represents matter. **We cannot, in this way, distinguish between mass and field, since the distinction between mass and energy is not a qualitative one. We could therefore say: Matter is where the concentration of energy is great, field where the concentration of energy is small. But if this is the case, then the difference between matter and field is a quantitative rather than a qualitative one.**" (emphasis by the author)
—A. Einstein and L. Infeld, *The Evolution of Physics*, Schuster, NY, (1961) (7)

The special theory of relativity also contained what he later saw as an error, his denial of the significance of the ether. It was not until 1919, three years after the publication of the General Theory of Relativity, in which he introduced his spacetime continuum, a four-dimensional construct of the three spatial dimensions with a fourth representing time, did he return to the notion of an ether, in a letter to Hendrik Lorentz:

"It would have been more correct if I had limited myself in my earlier publications (*referring to the Theory of Special Relativity*), to emphasizing only the nonexistence of an ether *velocity*, instead of arguing the total nonexistence of the ether, for I can see that with the word *ether* we say nothing else than that space has to be viewed as a carrier of physical qualities."
—A. Einstein, *Letter to H. A. Lorentz*, 15 November 1919, (8)

One finds no mention of an ether in the General Theory of Relativity. Only the terms space and time are there, But Einstein treats those abstract, non-physical entities as possessing real physical attributes, as if they can be curved, bent, truncated, manipulated as if they were the true clay of existence. His model is as abstract as Euclid's geometry, but based on the concepts of more modern mathematicians, Poincaré and Minkowski. At first appearance, it seems as if he has abandoned reality and replaced it with mathematics. It is only when one reads his utterances after 1919 that one begins to understand what he means when he says "space."

Realist though he claimed to be, Einstein loved his mathematics. He saw his equations as so beautiful they must be true. But mathematical beauty is deceiving. In the math everything is smooth, everything is continuous. It flows from the mind to the page, and gives the illusion of reality, a seduction to which the mathematical physicists still succumb today. But the world is not like that. The world is rough, coarse, vibrating. It is filled with turbulence, its edges grind and push against each other. In the Museum of Modern Art *mathematics* is a painting by Malevich, where the forms are clean, the edges crisp, the surfaces smooth, the juxtapositions measurable.

The *world*, on the other hand, is a Rothko. The forms are similar but the surfaces and edges extend into the empty space between, they mix, they crowd, they cry out. They are the Casimir effect illustrated.

Shortly after his letter to Lorentz, Einstein extended his thoughts about the ether. In a lecture delivered at the University of Leyden in May of 1920, he delivered his most complete defense of the concept.

"[...] in 1905, I was of the opinion that it was no longer allowed to speak about the ether in physics. This opinion, however, was too radical [...]. It does remain allowed, as always, to introduce a medium filling all space and to assume that *the electromagnetic fields (and matter as well) are its states*. But, it is not allowed to attribute to this medium a state of motion in each point, analog(ous) to ponderable matter. This ether may not be conceived as consisting of particles that can be individually tracked in time." [...].

"On the other hand there is a weighty argument to be adduced in favor of the ether hypothesis. To deny the existence of the ether means, in the last analysis, denying all physical properties to empty space. But such a view is inconsistent with the fundamental facts of mechanics."

31

"Summarizing, we can say that according to the theory of general relativity space is equipped with physical properties, also in this sense an ether exists. According to the general theory of relativity space without ether is unthinkable, as in such a space not only the propagation of light would not take place, but also there would be no possibility for clocks and rods, so that also no spatio-temporal distances "in the sense of physics.""
—A. Einstein, *Ether and Relativity Theory*, 5 May 1920, an der Reichs-Universität zu Leiden, Springer, Berlin (1920) (9)

Then, in 1930, he further published these thoughts:

[...] "the "aether of general relativity" is not absolute, because matter is influenced by the aether, just as matter influences the structure of the aether." [...]

"The real is conceived as a four-dimensional continuum with a unitary structure of a definite kind (metric and direction). The laws are differential equations, which with the structure mentioned satisfies, namely, the fields which support gravitation and electromagnetism. The material particles are positions of high density without singularity. We may summarize in symbolical language. Space, brought by the corporeal object, made a physical reality by Newton, has in the last few decades swallowed ether and time and seems about to swallow also the field and the corpuscles, so that it remains the sole medium of reality."
—A. Einstein, *Forum Philosophicum* 1, 180 (1930) (10)

But these were different times in physics. The 1920's saw the rise of the quantum theorists, and in small part, Einstein himself, with his vision of the character of light as being made up of "quanta" of energy. Working at the smallest scale of reality, quantum theory dominated the minds of most physicists as it still does today, along with a plethora of what many of us call 'speculative physics', with their emphasis on imaginary concepts with no apparent basis in the empiricist traditions of reasoning from observations in the real world. Thoughts about the ether receded into obscurity, so that the truth of Einstein's comments in 1930 seemed almost true again:

"[...] in 1905, I was of the opinion that it was no longer allowed to speak about the ether in physics. " (11)

But with Einstein's own help, the ether has returned, and it is time to take it seriously as a fundamental entity in physics once again. In 2009 a respected physicist, F. Selleri, published an extended history of the relationship of the ether to relativity and concluded that:

"Therefore one can say that "physical space" and "ether" are only different terms for indicating the same reality. Furthermore, fields are physical states of space. If no particular state of motion can be attributed to the ether, there does not seem to be any reason for introducing ether as an entity of a special type alongside of space. Naturally it is not forbidden to use the word ether, but only to express the physical properties of space."
—F. Selleri, Dipartimento di Fisica, Universita di Bari, "Relativistic physics from paradoxes to good sense -1", in *Ether, Space-time, and Cosmology*, Volume 2, 201-266, Duffy and Levy, Eds., Montreal (2009) (12)

So, if we accept, with Einstein, and now many other experts, that an ether must exist, else none of these other effects and phenomena could have occurred, what is its nature? What might it consist of? How might it work?

Where is this leading us? what is this train of thought leading up to? Let's pause here for a moment to consider what we are proposing. As you may have guessed, this book is based on the assumption that an ether of some sort exists, and not just as a container, but a medium that is also the source, the structure, and the generator of the universe which we inhabit. Our purpose in this part of that work is to define that ether, to show how it works, what it consists of, and how it manifests itself to us, not just in speculative theories and models, but in our everyday experience. It is our goal to demonstrate the high probability that this is a correct model of the universe and to try to modify the preconceived notions of the origins of the universe and its myriad parts and workings so that we might get past the many incompatible and contradictory theories that now pretend to explain it. One of those preconceived notions, is the conviction that all is "particles", like "the uncutable" "atoms" of Leucippus and Democritus of 5th century B.C. Greece, an idea that has governed our world view ever since.

Let's first look at this new, new ether (Einstein's was the first "new") and try to understand its qualitative nature. What are some the qualities "the ether," or by its other name, "space," must possess in order to carry out the multiplicity of functions that we know, by observation, it is capable of? What qualities must it have to be seen, as we suggested in our introduction, as the source material of our universe. Maybe first, what does it already do, in this real world we inhabit?

Primarily, here and now, it must be a medium that can carry the signals from this wireless keyboard on which I am typing these words to the internal mechanisms of this computer, and deliver them for display on the screen on which I can now read them. It must carry the light and the other electromagnetic energy through what we know as otherwise empty space to our exploratory vehicles on the moon and Mars, to monitor their behavior and progress, to give new instructions, to send back to us their findings. We know it supports radio waves, Radar, wireless communications of short wave signals, microwaves, heat energy from the sun, photosynthesis in nature, magnetic fields in motors, dynamos, and all the electronic devices we now, most of us at least, carry in our pockets. Did I mention the internet? It has carried the energy to earth that we now tap as carbon-based fuels, laid down millions of years ago, that we now find ourselves in danger of depleting. So something is there, detectable in many ways, that first created and supports our life on this planet. And it must exist well beyond the bounds of our atmosphere and our magnetosphere out in the near-vacuum of what we call space. We know it exists no matter what we call it. Let's choose ether.

How does it do these things? What physical characteristics must it have to manage all of the tasks that we know it does and so well? As we noted in the introduction, the kind of structure that most closely resembles what we have been describing is what we call an electromagnetic field. Fields have been known and studied since the early part of the nineteenth century, beginning with the work of Edmund Halley, who studied the earth's magnetic field, Hans Christian Oersted, who discovered in 1820 the connection between electricity and magnetism, and the work of André-Marie Ampére, Michael Faraday, and James Clerk Maxwell, whose four equations still guide our thinking about fields to this day. Wikipedia describes EM fields in this way:

"An electromagnetic field is a physical field produced by electrically charged objects. It affects the behavior of charged objects in the vicinity of the field. *(Once created)* **The electromagnetic field extends indefinitely throughout space** and describes the electromagnetic interaction.
The field can be viewed as the combination of an electric field and a magnetic field. The electric field is produced by stationary charges, and the magnetic field by moving charges (currents); these two are often described as the sources of the field. The way in which charges and currents interact with the electromagnetic field is described by Maxwell's equations and the Lorentz force law.
"From a classical perspective in the history of electromagnetism, the electromagnetic field can be regarded as a smooth, continuous field, propagated in a wavelike manner; whereas from the perspective of quantum field theory, the field is seen as quantized, being composed of individual particles. " (W) (emphasis added) (13)

Almost paradoxically, Maxwell's Equations suggest that an electrical field is created by a moving magnetic field and a magnetic field by a moving electrical field, sort of mushing up the mode of creation of both fields, or perhaps making the combination, the electromagnetic field, itself almost self-creating. This is what we here are positing as the composition of the ether, the boldfaced portions of the above citation intended to emphasize its unique character.

So here is where we stand, This is the first axiom of our argument. We cannot tell you how this ether actually came to be, but we are convinced that when you see how it meets the demands we make of it, the tasks it performs, the expectations it fulfills, that you will find it the most convincing, the most logical answer to how the universe we have come to know could have come about.

Axiom 1: Space is not a void or vacuum but is made up of energy in the form of a continuous electromagnetic ether, extending indefinitely, without limit, without edges or borders as far as the eye can see, even with the most powerful instruments. This field is fixed and is, in the relativistic sense, a privileged prime reference frame. It is, however, fluid and subject to the same internal movements, currents, turbulence, and topological defects as any elastic medium. This ether pervades all of space, as evidenced by its carriage of all electromagnetic phenomena to and from its farthest limits.

Here is how it might look:

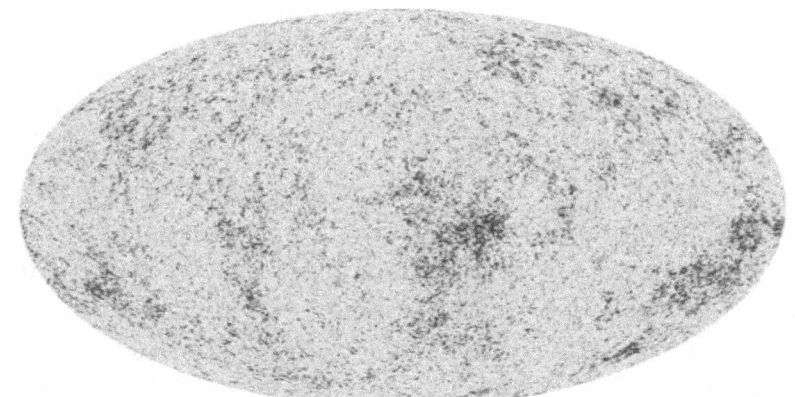

Image of the background radiation levels of the cosmos, by the Planck satellite. The color variation is computer generated temperature deviations with the darker portions about 2.5° Kelvin, up to the lighter portions, about 3° K. for an average temperature of 2.73° K. In the Big Bang theory of cosmology, this image is presumed to be the result of the primal explosion that created our universe. We argue that it is, in fact, the "boiling pot" out of which our universe and all that it contains is made up.

Let's examine what this means. How is this "ether" different from Boyle's or Newton's? What does it share with the ether of Hendrik Lorentz? Is this the ether that Einstein was referring to in 1920? Well, up to and including Lorentz's description in 1902, all concepts of the ether seemed to share one characteristic, a *particulate* nature. The fineness of its scale, how small its particles were, of course, was subject to the level of knowledge of the time. Certainly it was finer than could be detected by the naked eye, and it must be smaller than the entities it carried, light corpuscles for Newton, for example. Morley and Michelson saw it as being displaced by the passage of the earth, parted, in part carried along with it. Lorentz saw it as fine enough to pass through the interstices of the molecules that made up all matter, so that it was not displaced or deflected by any object's passage. This was his answer to Morley and Michelson as to why it did not slow light's passage. Einstein, in 1905, must either have been una-

ware of Lorentz's concept, or he felt it didn't matter whether
or not such a substance existed, or perhaps because Special
Relativity's principles didn't need an ether, one way or another. But by 1915, the time of his Theory of General Relativity, with its predictions and descriptions of the nature of gravity, forces and influences at great distances, the notion of a
substanceless, totally empty space must have seemed inade-
quate. We don't know of his speculations, but he must have
done some rethinking, else he would not have thought his letter to Lorentz was important. We also don't know whether he
was curious as to what the substance of this ether might be,
only that there must be "something out there," so that all
these other things might exist or happen.

Meanwhile the world went blithely along. The concepts of
the curvature of space and time became commonplace, unchallenged except for the few who might say, "I don't get that
but the genius of our time says it, so it must be so." And the
idea of an ether was, after all, pretty ethereal, and most of the
world loves mysteries. Besides that, another train of thought
began to grab the attention of physicists, the development of
quantum theory. We'll be coming back to that, but let's first
talk about how that field, our new, new ether, might be made
up.

1. What is it made of?

Two principal schools of thought have dominated the
world of theoretical physics over the last 350 years. *Classical
physics*, that is the physics of Leibniz and Newton on up
through all of the discoveries of the 18th and 19th centuries,
the development and discoveries of Gauss, Hertz, Faraday,
Lorentz, and James Clerk Maxwell in electricity and mag-

netism, even through Einstein's Special Theory of Relativity, dominated. In fact it was the only thing known until then. Einstein's General Theory of Relativity, replacing Newton's gravity with what he called *the curvature of spacetime*, initiated a revolution that might have displaced the classical theory completely had it lived up to its initial promise. General Relativity put in place not just spacetime as a concept, it replaced the smooth, continuous character of the universe, in gravity, in electromagnetism, all calculable in Leibniz and Newton's calculus, with what we now know as quantum theory.

Unwittingly started by Planck's invention of the quantum in 1900, picked up later by Einstein, a whole set of replacement theories for the classical model was being born. The notion that not only were some things in the universe not smooth, but that all things were ruled and must be considered in stepwise units called quanta, (whatever those were), made a new math and a new conceptual framework necessary. Remember that Max Planck's adoption of quanta (not called that by him, only named later) was what enabled him to describe the black box emissions of energy more accurately and precisely. He was not that thrilled that this notion was generalized to apply to everything, but it was enthusiastically adopted by other thinkers and mathematicians as the breakthrough kind of idea that opened all kinds of new doors for speculation.

What we saw from about 1913 on was the rise of these new "quantum theories" of how the world (and the universe) might be made. We already had electrons and protons and neutrons smaller than atoms themselves. Suddenly it seemed that there might well be building blocks even smaller than those. The first was the photon, the tiniest particle of light, finally becoming the energy unit of 1 quantum, temporarily. We are not sure, these days, whether that has changed and might be variable and not a constant, like light's velocity, but at any rate it was a place to start. The rise of particle theory, quantum theories, and the like is thoroughly documented elsewhere and will be discussed further on in this narrative, but for now let's see how it might have affected our concept of electromagnetic fields.

In Wikipedia's narrative called "Electromagnetic Field" the field's description is, to some extent equivocal. It is described in the first line of text in this way:

"An **electromagnetic field** (also **EMF** or **EM field**) is a physical field produced by electrically charged objects. It affects the behavior of charged objects in the vicinity of the field. The electromagnetic field extends indefinitely throughout space and describes the electromagnetic interaction. It is one of the four fundamental forces of nature (the others are gravitation, weak interaction and strong interaction)." (14) (W)

No definition is provided for "electrically charged objects." For this one must move to another subject heading. It further states that: One should also note that the term "interaction" is the currently favored usage for used to be called a *force* (Newton's term). Thus far the definition is in strictly "classical "terms, with one exception. The classification of the four "fundamental" interactions derives from particle and quantum theories, not classical physics. The description goes on:

"The field can be viewed as the combination of an electric field and a magnetic field. The electric field is produced by stationary charges, and the magnetic field by moving charges (currents); these two are often described as the sources of the field. The way in which charges and currents interact with the electromagnetic field is described by Maxwell's equations and the Lorentz force law."

In point of fact, Maxwell's Equations show that an electric field may be generated by a moving magnetic field and vice versa, so a moving "charged particle" is not a strictly required initiator. If one or the other finds itself present, the other must follow. Finally, Wikipedia makes itself clear:

"The electromagnetic field **may be viewed** in two distinct ways: a continuous structure or a discrete (quantum) structure." (15)

The first is the classical field of Faraday, Hertz, and Maxwell. The second may be thought of in a more 'coarse' way. Experiments reveal that in some circumstances electromagnetic energy transfer is better described as being carried in the form of packets called quanta (in this case, photons) with a fixed frequency.

The field we describe in Axiom 1 belongs to the first category, a classical field of continuous structure. This is our ether. This is the place, the location, the active participant, the source, of all we see, feel, taste, smell, hear, or detect with any combination of those senses or any prosthetic device that extends their range and sensitivity. What then are those objects, events, and phenomena made up of, and how? Is the current crop of physicists correct? Are they somehow made up of particles? And how are those made up? Or is there some other way to explain it?

1.3 Disappearing the Particle

First, a little diversion, in verse:

The Song of the Simple Universe
(Hadrons are a class of fundamental particles enshrined in the lexicon of the standard model of modern physics. The Simple Universe is the theory that all is energy.)

Who has seen a hadron?
Who knows she exists?
Clothed in colors, flavors, spins.
Wrinkles in a mist.

Who has seen a hadron?
Called her by her name?
Fermi, bose, neutrino, higgs,
Are they not the same?

We have seen the hadrons.
Cosmic wrinkles small,
Distant giants, moons and stars,
Creatures short and tall,
Walked among them, tasted, touched,
Energetic, all.

(To be sung to the tune of "Who is Sylvia?" words by William Shakespeare, melody by Franz Schubert)
—Charles Scurlock, 2013

imagine darkness

1. *The Uncutables*

The concept that matter is made up of tiny particles is a notion that has a long history. As we mentioned earlier, Leucippus and his pupil Democritus, in the fifth century BC, put
forward the idea that if an object, consisting of matter, were
cut in two, and then cut in two again, and again, that eventually one would reach a size that would be "uncutable", hence
we call that fundamental entity an "atom," the Greek word
for uncutable. Remember, these early thinkers could see dust-
motes, perhaps the smallest detectable entities in their universe, so the act of imagining something even smaller was a
great intellectual leap into the unknown. But they clearly
shared the quality of thinking with us that I have come to call
the contemplative mind, that of taking in facts and observations and wondering, "How could this have come about?"
Might there be something even smaller?" "And what might
that look like?" Epicurus followed with elaborations of the
same idea, a hundred years later, but most of his writings
have been lost.

Then came Lucretius, in the first century BC, who, giving
full credit to Epicurus, expanded the concept of what he came
to call "first-beginnings" in great detail, how they must be of
many sizes and shapes, how they might fit together and adhere so as to form larger and more complex structures, even
the everyday things we see around us.

Lucretius' wonderful justification for his view of creation
came out of an argument that nothing can come from nothing, that, for example, maggots could not spontaneously arise
from dead flesh, but even further:

"For if of all things came out of nothing all kinds of things could be
produced from all things, nothing would want a seed. Firstly man
could arise from the sea, from the earth scaly tribes, and birds

could hatch from the sky, cattle and other farm animals and every kind of wild creature wood fill Desert uncultivated land alike with no certainty as to birth.

But as it is, because every kind is produced from fixed seeds, the source of everything that is born and comes forth into the borders of light is that in is that in which is the material of it and its first bodies; and therefore it is impossible that all things be born from all things because in particular things resides a distinct power."

    —Lucretius, *On the Nature of Things*, 1, 157-184,
Loeb Classical Library, Harvard (1)

And further:

"Add to this that nature resolves everything again into its elements, and does not reduce things to nothing. For if anything were perishable in all its parts each thing would then perish in a moment snatched away from our sight. For there would be no need of any force to cause disruption of it's parts and dissolve their connections. But as it is, because this because the seed of all things is everlasting nature allows no destruction of anything to be seen, until a force has met it sufficient to shatter it with a blow, or to penetrate within through the void places and break it up.

"Whence is the sea supplied by the springs within it, and by the rivers without flowing from afar? Whence does the ether nourish the stars? For all things that are of perishable body must have been consumed by infinite time and ages past. But if through that space of time past there have been bodies from which the sum of things subsists being made again, imperishable indeed must their nature be; therefore things cannot severally return to nothing,"
—Lucretius 1, (213-239) (2)

After describing all manner of things, He finally feels he has made his point, that all things must proceed from something, and lays out for us what that something, those "first-beginnings," must be.

"Yet everything is not held close and packed everywhere in one

45

solid mass, for there is void in things: which knowledge will be useful to you in many matters and will not allow you to wander in doubt and always to be at a loss as regards the universe and to distrust my words. Therefore there is intangible space, void, emptiness. But if there were none, things could not in any way move; For that which is the province of body, to prevent and to obstruct, would at all times be present to all things; therefore nothing would be able to move forward, since nothing could begin to give place. But as it is we discern before our eyes throughout seas and lands and the heights of heaven many things moving in many ways and various manners, which if there were no void, would not so much lack altogether their restless motion, as never would've been anyway produced at all since matter would have been everywhere quiescent packed in one solid mass."
—Lucretius, 1 (329-347) (3)

"Then further if there were nothing void and empty, the universe would be solid unless on the other hand there were definite bodies to fill up the places they held and the existing universe would be vacant and empty space therefore without doubt. Body is marked off from void alternately since the universe is not completely full nor yet empty. There are therefore definite bodies to mark off empty space from full. These can either be dissolved by blows when struck from without nor again when pierced inwardly and decomposed, nor can they be assailed or shaken in any other way as I've shown you above a little while ago. For it is seen that without void nothing can be crushed or broken or split in two by cutting, nothing can admit liquid or again percolating cold or penetrating fire, by which all things are destroyed. And the more each thing holds void within it so much the more thoroughly it is shaken when these things attack it. Therefore if the first bodies are solid and without void, as I have taught, these must be everlasting. "
"Besides, unless matter had been everlasting, before this all things would have returned utterly to nothing, and whatever we see would have been born again from nothing. But since I have shown above that nothing can be produced from nothing and what has been made cannot be brought back to nothing, there must be first beginnings of immortal body, into which each thing can be

resolved at its last moment, that matter may be forthcoming for the renewal of things. The first-beginnings are therefore of solid singleness nor can they in any other way be preserved through the ages from infinite time passed and make things anew."
—Lucretius 1 (520-548) (4)

These copious quotations from Lucretius, from 2000 years ago, have a purpose. His work, these six books, was presumed lost from about the second century AD until they were rediscovered 1200 years later, stored in the archives of a small monastery in southern Germany. Their publication, many scholars believe, was a powerful force in what became known as "the enlightenment," that great surge in scholarship that led to the work of many of our great scientists of the 16th, 17th, and 18th centuries, among them Copernicus, Kepler, Galileo, and Newton and to a new and more rational understanding of the world and how to think about it. They also show how, in the absence of microscopes, telescopes, and other modern aids to our researches and observations, that Lucretius' ideas and theories, and his model of the universe, held the minds of scientists in thrall for another 400 years.

There are two important concepts there in the quotations from Lucretius. One is that the universe contains, and in fact, must contain, both matter and void. Without either, all would fail. We could not have the universe we perceive. The other is time-bound, that everything that constitutes matter is made up of atoms, something that cannot be further cut in two, his "first-beginnings." And these connect together in various ways to make up the vast array of matter-entities in the world around us. (note also that he believed in an ether!)

While the idea of atoms as the fundamental constituent of matter had a long history, it was an esoteric theory, known and communicated among a few. And as we know, Lucretius' accounting of it was lost for hundreds of years. The commonsense theory was that all of creation consisted of just four elements that together contributed their qualities to the world. These were earth, air, fire, and water. Aristotle described their qualities as:

Fire is primarily hot and secondarily dry.
Air is primarily wet and secondarily hot.
Water is primarily cold and secondarily wet.
Earth is primarily dry and secondarily cold.

This was not something on which science had much to say, although it affected much of general thinking including medicine, religious beliefs and the like.

Atoms as the fundament of matter were revived in modern times in 1661 when Robert Boyle published *The Sceptical Chymist* (5) in which he argued that matter was composed of various combinations of different "corpuscules" or atoms, rather than the classical elements. During the 1670's corpuscularianism was used by Isaac Newton in his development of the corpuscular theory of light.

In 1789 Antoine Lavoisier described the law of conservation of mass and defined an element as a basic substance that could not be further broken down by the methods of chemistry. (6)

In 1805, English instructor and natural philosopher John Dalton proposed that each element consists of atoms of a single, unique type, and that these atoms can join together to form chemical compounds.(7) Dalton is considered the originator of modern atomic theory. Dalton's atomic hypothesis did not specify the size of atoms. Common sense indicated they must be very small, but nobody knew how small. Therefore it was a major landmark when in 1865 Johann Josef

Loschmidt measured the size of the molecules that make up air. (8)

In 1827 botanist Robert Brown used a microscope to look at dust grains floating in water and discovered that they moved about erratically—a phenomenon that became known as "Brownian motion," (9) and in 1905 Albert Einstein produced the first mathematical analysis of the motion. French physicist Jean Perrin used Einstein's work to experimentally determine the mass and dimensions of atoms, thereby conclusively verifying Dalton's atomic theory. Note that no one had yet seen an atom. The mathematics was their only tool to determine what they assumed to be the size of an atom. In later times, we will see, physicists came to depend on energy measurements to estimate the size and mass of their "elementary" particles.

The physicist J. J. Thomson, through his work on cathode rays in 1897, discovered the electron, and concluded that it was a component of every atom. Thus he overturned the belief that atoms are the indivisible, ultimate particles of matter. Thomson postulated that the low mass, negatively charged electrons were distributed throughout the atom, possibly rotating in rings, with their charge balanced by the presence of a uniform sea of positive charge. This later became known as the plum pudding model. (10)

In 1913, physicist Niels Bohr suggested that the electrons could best be seen as being confined into clearly defined, quantized orbits. They could jump between these orbits, but could not freely spiral inward or outward in intermediate states. An electron must absorb or emit specific amounts of energy to transition between these fixed orbits. This Bohr model of the structure of the atom became the accepted standard and was used to explain the concepts of ionization and multiple isotopes of the same element, as well as the mechanism by which unrelated atoms could connect with others to form chemical compounds. The Bohr model of the atom is a clear and simple one. Electrons orbit a nucleus in one or more definite orbits at different distance from the center. A specific level of energy is required for an electron to move from one orbit to another. These are designated as quanta of energy, hence the origin of the phrase "quantum leap." (11)

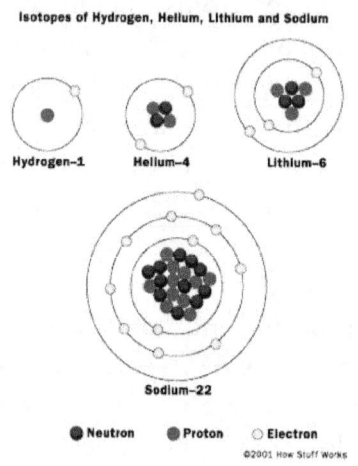

Isotopes of Hydrogen, Helium, Lithium and Sodium

Hydrogen–1   Helium–4   Lithium–6

Sodium–22

● Neutron   ● Proton   ○ Electron
©2001 How Stuff Works

The numbers shown are what is called the atomic weight of the element, typically the sum of protons and neutrons making up the nucleus, electrons making negligible contributions to the mass of an atom, since they carry a mass that is less than 1/1800th of their protons and neutrons. (from Wikipedia, 2014) Subatomic particles

"Though the word *atom* originally denoted a particle that cannot be cut into smaller particles, in modern scientific usage the atom is composed of various subatomic particles. The constituent particles of an atom are the electron, the proton and the neutron; all three are fermions. However, the hydrogen-1 atom has no neutrons and the hydrogen ion has no electrons.
The electron is by far the least massive of these particles at 9.11×10^{-31} kg, with a negative electrical charge and a size that is too small to be measured using available techniques. It is the

lightest particle with a positive rest mass. [. . .]
Protons have a positive charge and a mass 1,836 times that of the electron, at 1.6726×10^{-27} kg. The number of protons in an atom is called its atomic number. [. . .]
Neutrons have no electrical charge and have a free mass of 1,839 times the mass of the electron. [. . .]The neutron was discovered in 1932 by the English physicist James Chadwick.
In the Standard Model of physics, electrons are truly elementary particles with no internal structure. However, both protons and neutrons are composite particles composed of elementary particles called quarks. There are two types of quarks in atoms, each having a fractional electric charge. **Protons are composed of two up quarks (each with charge $+\frac{2}{3}$) and one down quark (with a charge of $-\frac{1}{3}$). Neutrons consist of one up quark and two down quarks. This distinction accounts for the difference in mass and charge between the two particles.** (Wikipedia) (emphasis added) (12)

To make this model even more complex, the forces which hold these particles together must be defined, which requires the invention of even more particles: (Author's Note: The following paragraph is presented exactly as it appears in Wikipedia. We are not clued in as to what form a "boson" takes, what a "gluon" is, what the term "mediate" implies, or what constitutes a "nuclear force," except that it appears to be a "residuum" of some other "strong" force."

"The quarks are held together by the strong interaction (or strong force), which is mediated by gluons. The protons and neutrons, in turn, are held to each other in the nucleus by the nuclear force, which is a residuum of the strong force that has somewhat different range-properties (see the article on the nuclear force for more). The gluon is a member of the family of gauge bosons, which are elementary particles that mediate physical forces. "(Wikipedia) (13)

In 1964, Murray Gell-Mann posited quarks as the constituent parts of both protons and neutrons to explain away discrepancies in the mass and charge of the two particles. Like many parts of the quantum lexicon, the name "quark" has nothing to do with reality. It comes from James Joyce's novel Finnegan's Wake, from the line, "Three quarks for Muster Mark!" Other imagined qualities of fundamental particles are colors, flavors, spin, and strangeness, none of which denote anything approaching the common meaning of those terms. While quarks are posited to make up the composition of protons and neutrons, they cannot be identified outside of these composite structures. In a sense, they are conceptual only, there to explain both the charge, or not, of the two fundamental particles, or to explain an otherwise unexplainable result of the detection process in colliders where they appear as otherwise unidentifiable products of particle collisions.

How many fundamental particles are there? Well, the number continues to grow, as more are seen as necessary to explain, or explain away, contradictory or unknown outcomes of experiments. The family of so-called "fundamental" particles now numbers 61. But wait, in order to explain other complications in the data such as the sudden disappearance of a particle or other unknown effect, there is something called an antiparticle, which cancels out the existence of the original one. These have not yet been seen either. There are even "virtual" particles, those whose existence is so brief that they are undetectable.

How do we "know" that these "fundamental" constituents of nature are "particles"? Is it because Democritus and then Lucretius called them that? Is it because the idea is so ingrained in our mental models of the world that we cannot imagine it any other way? As we mentioned earlier, Newton imagined light as made up of "corpuscles", individual units, though very small, to explain how refraction of light occurred.

Then of course, others showed, for about the next 200 years that light must be energy in the form of waves, which satisfied physicists until Max Planck and then Einstein shifted us back to the corpuscular side with their "quanta." Another reason probably lies in how they are detected. In the Pacific northwest there is a persistent myth of a mysterious forest figure called Sasquatch, humanoid, but hairy and giant, living deep in the wilderness, but never having been actually sighted or having its actual existence confirmed. When a true believer is asked for evidence of its existence, the usual response is, "Well, I haven't seen one myself but I know someone (or a friend of my second cousin does, or some anthropologist) who has seen its tracks."

In modern physics the way we know about these things too small for detection by the naked eye is how we have always known them. We know there is air around us because another sense, not vision, can sense the wind, or the heat or the cold, and though we cannot see the agent, we know it is there. We observe fire, explosions, the movements of planets, we measure forces and velocities, and from these observations we can say that something has caused these movements, these appearances, something that must exist although we can't see it. This is perfectly reasonable. It works, and is the basis for a huge part of the enormous store of knowledge we have accumulated about the world around us, even the universe. But...

2. Who has seen a hadron?

For detecting particles, we use highly sophisticated equipment that we call, of course, detectors. John W. Moffatt, in his admirable book, *Cracking the Particle Code of the Universe*, (14) devotes a long and thorough chapter to the tools and means of detection of the "particles" that are his principal subject.

You may recall, we earlier referenced the "revolution in physics" triggered by Planck and Einstein in 1900-1905, the shift to "quantization." Moffatt describes that revolution this way:

"This was a complete surprise in experimental physics at the time. It was widely accepted and had been proved by James Clerk Maxwell back in the mid-19th century that light existed as waves. In fact, Planck never accepted the idea that his radical discovery of packets of energy was true and described nature accurately, nor did many of his colleagues. Einstein, the iconoclast, took Planck's discovery a step further in 1905 by claiming not only that the blackbody walls produced light in quantum packages but that light itself consisted of quantum packages in the form of particles which we now call *photons.*" (15)

The term "quantization" can, of course, be looked at in more than one way. As accepted and used by most modern physicists, it means that energy, light in particular, exists only as tiny capsules or packets, separate and unique from others in its vicinity, each carrying a specific unit of energy. On the other hand, "quantization" can mean that there is a smallest unit of measurement that we can use to describe the characteristic size or volume or magnitude of an entity. For example, if I hold up to you a 1-liter bottle of water I can tell you that this bottle contains 1000 cubic centimeters of water. I have "quantized" that liter of liquid. But if I then pour that liter of water out onto the ground, it doesn't flow out in little 1-cc units, it flows out continuously. What Max Planck was attempting to do was to make his calculations work out. When he found that he could use a very small unit, and not a smaller one, he called that unit a "quantum" of energy. He did not "discover" the quantum, he invented it to make the math work. No wonder he was skeptical when the "quantum" became an *ipso facto* "real" object in his colleagues' minds.

It should also be pointed out that many phenomena, but particularly those that appear to us as waves, which is the analogy most used in describing energy, are by definition alternatingly up or down, varying in intensity, positive or negative; and most important, that they may go in and out of a threshold of perceptibility. If, for example, one's mechanism of detection can only read the "positive" level of a wave-like impulse, if one can only "see" the parts above that threshold and not those below it, the resultant output from our detector will appear as separated impulses, especially when we are dealing with impulses as small as 10^{-35}, the scale of Planck's Constant. No, the adoption of the quantum as a separate and distinct physical unit in physics was a matter of convenience to the mathematical minded, not a great "discovery." (We will return to the argument of "discoveries" vs. "inventions" further down the line.)

The idea that energy only manifested itself in discrete units, as of photons of light, was resisted for a time, even though it was consistent with the "particle "mindset that dominated all concepts of the makeup of matter. "If matter (obviously) is made up of particles, why shouldn't energy be as well?" The particle nature of light and with it all other electromagnetic energy was further supported by the experiments of Compton, Cherenkov and others. Numerous effects, those of a unit of charge (an electron) passing through an electric field, or though a supersaturated container of water vapor (Wilson's cloud chamber), Einstein's "Photoelectric Effect" for which he received his only Nobel award, lent more credence to the idea of a particle nature for all radiation. The idea that a continuous phenomenon might in some instances only be seen as separate units was never considered.

Still, so far, no one has yet *seen* a hadron, only her tracks.

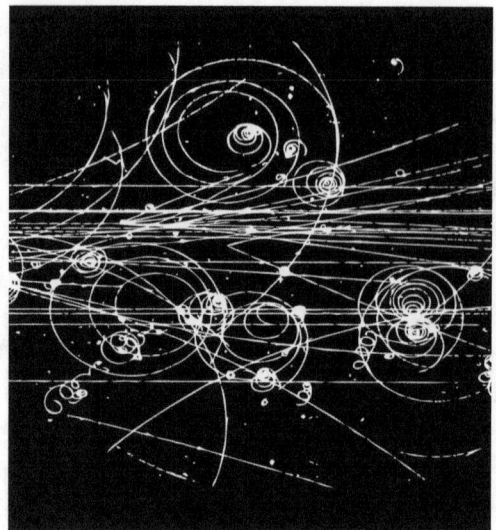

The picture shows the interactions in a particle stream in a bubble chamber filled with liquid hydrogen. Bubble chambers help to detect weakly ionized molecules in the solution, some, the spiraling ones, strongly affected by application of a magnetic field.

But even as Einstein, and later his colleagues (and rivals) in the physics community proceeded along the "quantum path" the adoption of the concept was fraught with problems, most particularly, the paradox of what has come to be known as the "Double-slit Experiment."
(from Wikipedia)

"While studying medicine at Göttingen in the 1790s, Thomas Young wrote a thesis on the physical and mathematical properties of sound[3] and in 1799, he presented a paper to the Royal Society where he argued that light was also a wave motion. His idea was furiously opposed because it contradicted Newton, whose views were considered sacred. (W)Nonetheless, he continued to develop his ideas. He believed that a wave model could much better explain many aspects of light propagation than the corpuscular model:
"A very extensive class of phenomena leads us still more directly

to the same conclusion; they consist chiefly of the production of colours by means of transparent plates, and by diffraction or inflection, none of which have been explained upon the supposition of emanation, in a manner sufficiently minute or comprehensive to satisfy the most candid even of the advocates for the projectile system; while on the other hand all of them may be at once understood, from the effect of the interference of double lights, in a manner nearly similar to that which constitutes in sound the sensation of a beat, when two strings forming an imperfect unison, are heard to vibrate together."

He also demonstrated the phenomenon of interference in water waves.

In 1801, he presented a famous paper to the Royal Society entitled "On the Theory of Light and Colours" which described various interference phenomena, and in 1803 he performed his famous double-slit experiment (W) (15)

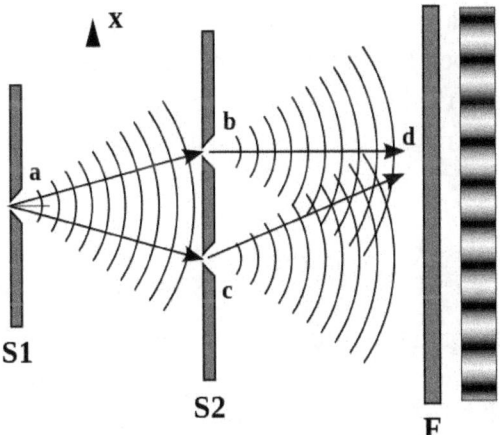

Light radiating from point a passes through two narrow slits at points b and c. The pattern resulting from the interference of waves from the two slits results in the pattern displayed on the detector plate at F.

The conflict is obvious. This is a repeatable experiment,

and it has been repeated hundreds of times. It demonstrates a phenomenon not explainable by conceiving of light as being made up of "particles." Yet to this day, the particle theorists insist that this is the case. With the rise of a new theory in the 1920's and thereafter, called "Quantum Theory," the need became obvious that a way out of this dilemma was necessary. Without any evidence at all to support it, the answer became, "Well, of course, waves can be sometimes behave as particles, particles can behave as waves. And there is no way to tell by observation which behavior can be predicted. We will call this phenomenon "wave-particle duality" and let it go at that."

When pressed for details, the explanation became that a particle of course carries with it something called a "wave-function" which at some point in its existence, "collapses" and the particle nature of the entity becomes what we observe. Whole books have been written on this subject alone, listed in our References. Let us just say for now that the answers as to how these experimentally unconfirmable models gained ascendance appears to be 1, through the objectification of mathematical models, and 2, a tendency toward eastern mysticism on the part of many of its protagonists. (16)

3. Wave-particle Duality Reconsidered.

The so-called wave-particle duality, of course depends on the assumption that matter is made up of particles, and that since is transformable into energy ($E=mc^2$, after all), it can Easily take the form of energy, This without any mechanics of transformation, just by the magical process of passing through a slit!. Or, on the other hand, by the act of observing it , we by also some magical process, flip it back and forth between these two states. I used to ask the question, about this quasi-

Berkeleyian notion, "Do you mean that this transformation could not have happened in the eons before we, as observers, inhabited the earth?" In the course of asking these questions I happened to find the following discussion by a blogger who carefully guards his real identity, but which makes too much sense to ignore.

(from lgsims96)
"Waves can exhibit particle-like characteristics and particles can exhibit wave-like characteristics. Is it possible that waves alone could behave as particles? What if the universe is composed of a single medium capable of supporting vibrations? And if all that we perceive as matter and energy is only vibrations (electromagnetic waves) within this medium? This is all there is and nothing else.
Years ago the ether was proposed as a medium to support the movement of electromagnetic waves through empty space. After all you can not have water waves if you have no water to support the waves. You can not have sound waves without air or some other medium to support the wave.
One major objection to the idea of the ether is that it would cause resistance to the movement of matter through the medium. That objection disappears if matter itself is only a vibration in the medium. Thus, without the medium there is no light, there is no matter. It might be this medium is solid. It certainly must have a high rigidity to transport light waves at such a high velocity.
With this in mind, we can use imagination, to suggest possible explanations for some observed physical phenomenon. How could the combinations of various electromagnetic waves or impulses ever behave as a particle? Let us compare attributes of particles with those of waves.

Attributes of particles
A particle has mass, it is localized in space. Two or more particles cannot occupy the same space at the same time. A particle can have any relative velocity from 0 to almost c (the speed of light).

Attributes of waves
An electromagnetic (EM) wave has no mass. It is not localized; it spreads out over a large volume of space. Many waves can occupy the same space at the same time. These waves have only one relative velocity c. They have attributes of wavelength, frequency, intensity and amplitude of the disturbance (electric charge)."
—lgsims96, *HubPages*, worldwide web, (March 2013) (17)

lgsims96 is a username on HubPages, an internet posting and discussion site. He is by far not alone in raising this question. A few examples:

Michael Faraday
"I have long held an opinion almost amounting to conviction. . . that the various forms under which **the forces of matter are made manifest have one common origin**, or in other words, are so directly related and mutually dependent that they are convertible, as it were, one into another, and possess equivalents of power in their action." (18)
—Michael Faraday, *Effects of Magnetism on Light*, 1845

Hendrik A. Lorentz
"[....] **and this leads us to the idea that an atom is in the last resort some sort of local modification of the omnipresent ether, a modification which can shift from place to place without the medium itself altering its position.** Having reached this point, we can consider the ether as a substance of a completely distinctive nature, completely different from all ponderable matter." (19)
—Hendrik Lorentz, Nobel Prize acceptance lecture, 1902

Albert Einstein
$E=mc^2$ (20)
—A. Einstein, *On the Electrodynamics of Moving Bodies*, (1905)

Einstein was halfway there with his lecture at Leiden University on May 5, 1920, when he said:

Albert Einstein
"Since according to our present conceptions **the elementary particles of matter are also, in their essence, nothing else than condensations of the electromagnetic field,** our present view of the universe presents two realities which are completely separated from each other conceptually, although connected causally, namely, gravitational ether and electromagnetic field, or as they might also be called space and matter." (21)

Alexander Unzicker
"For the entire 19th-century continuum mechanics was believed to be a valuable description of electrodynamics. Physicists imagined electromagnetic waves as propagating oscillations of an elastic medium called the ether which was believed to permeate all of space."
The ether theories vanished after 1905, because Einstein's theory of relativity didn't need ether and the experimenters at that time couldn't find it. In fact the idea of masses gliding through the ether as fish do through water leads to contradictions.
However, there is an intriguing analogy. **Wave structures and other irregularities in an elastic continuum (defects)** surprisingly behave like particles. Being nothing but the nucleus of a disturbance it cannot move faster than the disturbance itself, and *(in the atmosphere, for example)* the motion of the disturbance is limited by the speed of sound. The analogy to electron motion which is limited by the speed of light is obvious. **Moreover the formulas for charged particles in the special theory of relativity are identical to those describing the motion of defects in elastic solids.** This is exciting because it could mean that the ether was abandoned prematurely since people didn't know about the possibility of modeling particles as defects in such an elastic solid."
—Alexander Unzicker, *Bankrupting Physics*, Springer-Verlag, Heidelberg, Pangrave Macmillan, New York (2013) (22)

These, of course, are not all of the thinkers who imagined a different structure for what we call matter, different from be-

ing made up of tiny, uncutable elementary units, "first beginnings", "atoms", "particles," so to speak. These fundamental
entities are, from above:

"A **local modification of the omnipresent ether**" (Lorentz),
"**condensations of the electromagnetic field**" (Einstein)
"**Wave structures and other irregularities (defects) in an
elastic continuum**" (Unzicker)

So, if matter is not made up of particles, what might it be
made up of?

It is an easy step from these generalizations to the notion
that a perceived "particle" might be portion of such a defect,
or condensation, or ether modification, one that goes in and
out of our boundaries of perception, giving the appearance of
a series of separate entities, not the continuous wave that is
the phenomenon's actual nature.

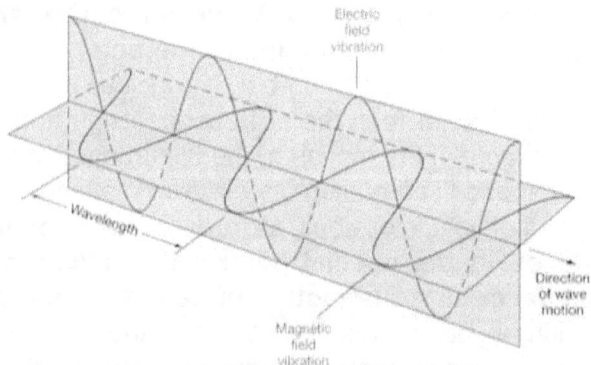

Schematic representation of an electromagnetic wave. In reality such waves in an electromagnetic field are more likely to
be vibrating impulses like hollow spheres of charge of varying
intensity than being two waves at right angles to one another.

Or it might be simply that what we perceive as particles are
coherent, organized (temporarily stable) foci of energy, but
still part of the ether, the medium we see as the source of eve-

rything. "Temporary" on a cosmic scale might, of course, be billions of years.

Let's look again at Unzicker's comment. *"Wave structures and other irregularities (defects) in an elastic continuum surprisingly behave like particles."* He suggests a direct analogy might be that of sound traveling in an elastic medium such as air or water. If one considers a normal atmosphere like ours, sound waves, which are, in fact, compression waves in the medium of the air, travel at a constant velocity as long as the medium is of uniform pressure and density. Generating a sound creates a sequence of higher and lower pressure vibrations in the medium. Note here that sound waves are not made up of some esoteric material or bundles of particles (sonitons, anyone?) passing *through* the medium. Rather they are actually structural distortions of the medium itself. The velocity of a sound wave, about 1100 fps at sea level, is not determined by the energy (loudness) or the frequency (a particular pitch), but are inherent in the (relatively) constant medium of which they are an integral part.. A low pitched, soft note travels at the same velocity as a loud, high pitched one.

And most sounds are not pure. Even a single tone from say, a flute, contains not just one set of vibrations, but many. These can be attributed to the material of the instrument, wood or metal, sometimes called tone color; as well as overtones, that is, a mix of a fundamental pitches and resonant higher frequencies produced by sympathetic vibrations in the materials of the instrument. If one examines the wave forms emitted by the instrument, one finds a mix of complex, overlapping patterns that can make the sound from a particular flute unique and identifiable to a person with a well-trained ear. Now multiply the sources of the sound.

But to make this example even more complex, our flautist sits in the third row of an ensemble of 100 musicians. There are 3 more flutes, of course, each slightly different from the others, so that the sound carries slightly more complexity, but the ninety-six other players are creating related sounds on different instruments, each of which disturbs the atmosphere in a different way. So, if you attempted to analyze the makeup of the many hundreds of soundwaves reaching you in the third row of the second balcony, you would find it extremely difficult. On the other hand, as a whole, the sounds and tone colors and resonances and harmonies reaching your ears comprise a whole, powerful experience as only a performance of Mahler's Ninth Symphony can do.

If we start to analyze the individual parts of the experience what we find is that each tone and its set of overtones from each flute, violin, horn in the ensemble has generated its sound on the basis of some simple rules. The vibration of a bow drawn across a string of metal or gut; the vibration of the lips of the trumpet player passing through and absorbing the sympathetic vibrations of the horn; a reed vibrating in the lips of the bassoonist. Each combination, say the output of 3 bass violins, is made up of several patterns of frequencies and amplitudes of the complex compression waves in the atmosphere of the auditorium. And each of these in their sonorous relation with other instruments, interacts with and modifies the output that reaches your ears.

All depends on the temperature, pressure and density of the medium and the way it behaves when vibrated. All of the different sounds reached you simultaneously, because for the most part they emanated from a single somewhat diffuse point on the stage, implying that the velocity of sound from the string bass was exactly the same as that from the piccolo or the percussionist's triangle. And you will understand that this magnificent, rich, complex experience resulted from a small set of simple rules, inherent in the medium, and manipulated by the players. The rules will have to do with what we know about phenomena like reverberation, reinforcement, resonance, which, if in opposition, damps sounds almost to silence, but which in concert magnifies the sound almost to too great an intensity.

So, what is sound? It is a distortion of its medium such that a perceptible difference is sensed. What is music? Also a distortion in the medium, but an ordered, coherent one that conveys distinct patterns to the listener. The first is noise, the second carries information. But note that this information is at the next level of abstraction from the complex set of vibrations reaching your ears.

Another example: From our experience and observations here in our ordinary experience, in what I have called "the zone of middle dimensions," the place where, in general, Newton's laws are enough, we know how our familiar radio-wave phenomena are propagated, that is, carried, from a broadcast base to our receiving devices, This is accomplished by imposing the information-carrying signal on a so-called "carrier" wave at a specific average frequency. For example, here is how modern commercial radio signals can be described.

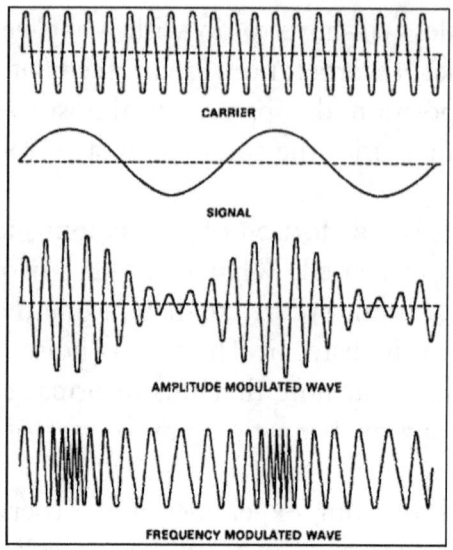

In radio and in other forms of electromagnetic communication, all of the complex information, all of the sounds, colors, actions being broadcast are first transformed into electromagnetic waves, then sent out as part of a carrier wave, then converted back into the radio and television sounds and images that we enjoy in our living rooms.

What we are saying is that there exists as the ether at least one and perhaps a range of fundamental frequencies that serve as the base, the carrier, for all other electromagnetic impulses. That range of phenomena contains every level of perceptible energy in the universe, as well as those not yet perceived by us or our instruments. If the ether carries this wide range of frequencies throughout the universe, it must function at a higher frequency than those we have detected thus far. The highest frequencies detected until now are what we call gamma rays in the range of 10^{30}, that is, 10 with 30 zeros following it. So for our electromagnetic ether to be the carrier of the full range of what we designate as the detectable electromagnetic spectrum, its native frequency, its high energy, must be beyond that 10^{30} level that we have detected. For example, its frequency might lie in the range of $1/h$, that is, 1 divided by Planck's Constant, about 1.15×10^{35}, a very high frequency.

What can also be seen, in the graphic of the radio wave example, is that the "signal" portion of the waveforms shown, does not "ride" on or pass through the carrier wave. What it does, in fact, is to effect a physical distortion of the original carrier wave. That is, there are not two distinct waves emerging from the imposition of one onto the other, but a change in state of the original, carrier" wave, creating a new and distinct form. But both the original wave and the new wave are made up of the same physical substance, energy in its electromagnetic form.

In the early 1920's a young electrical engineer named Balthasar van der Pol was studying the effects of oscillating electromagnetic frequencies by varying the intensity of electrical currents to an early vacuum tube. Without an oscilloscope to help him see and perhaps measure the effects, he listened to tones generated from the vacuum tube through an ordinary telephone handset. As he increased the current directed to the vacuum tube, he found that the sound in his handset first passed through static, what we now would call noise, but at certain current intensities it would generate a coherent sound at a particular pitch,. As he increased the level of energy into the system, this alternation of noise and coherent sound repeated itself at higher pitches. Order arose out of chaos by the application of energy. In the noisy bands, he was hearing random, conflicting patterns in the sound. In the coherent sounds, he was receiving patterns of reinforcement and resonance, In other words, order appeared out of chaos by the increasing the energy of the (local) system. This is but one mechanism demonstrating that orderly, particle-like behavior can appear out of otherwise turbulent, even chaotic systems. We will show in Part 3 some of the other mechanisms that can result in stable, coherent entities, without the need for the existence of uncutable primary constituents. (23)

At the risk of reader boredom, I will repeat, "No one has seen a 'particle." Researchers have recorded seeming points on detector systems, they have heard clicks on Geiger counters, seen tracks in cloud and bubble chambers, have measured energy levels in collider systems like the Large Hadron Collider at CERN. They have accepted and proven that $E=mc^2$ and denote their particles' masses in millions of "ev", (electron volts). They have developed a massive taxonomy of "fundamental particles" (61), and their opposite number of "antiparticles." A study of the literature and of elementary logic tells us that such a panoply of "fundamental" particles, none of them ever seen or their effects perceived, many just presumed to exist because the math requires it; that exist for the most part as only symbols in mathematical equations, is not clear enough justification for a claim for their existence. On the other hand, we are convinced that this huge, complex model of the makeup of the real world is the result of observations biased through misattribution, of being assumed to have one obvious cause, passage of a particle through a medium; rather than another obvious cause, distortion of the medium itself resulting in perceptible wrinkles in that three-dimensional fabric.

Particle physics and its theoretical basis, quantum mechanics, with their multiple contradictions and logical paradoxes is simply not sufficient to describe physical reality.

If we then adopt the wave theorists' position and call "particles" what they really are, organized, coherent concentrations of energy, the paradoxes of *superposition, wave-particle duality, wavefunction,* and the like, simply disappear, as does the need for a label such as *particle.* RIP. Some will weep at the supposed loss of mystery, but we should not weep with them. There will always be a place for mysteries in science, because the farther we dig into its structure the more we find. The next thing that we realize is that, when we place this new structure into the ether we find that it is right at home, since it is both within and a part of its environment, just as sound waves are part of the atmosphere, similarly coherent, organized, and temporary, though on a much different time scale.

4. The Smallest Parts

Axiom 2. There are no particles, no uncutable first-beginnings, no atoms, no protons, neutrons, electrons, quarks, neutrinos, bosons, gluons; again, no "particles." There is only energy, flowing through and within the cosmos. Its formation into concentrations of various size, intensity, and complexity gives rise to the illusion of entities of a particulate nature.

Its distribution is random, not ordered. Its smoothness is nearly but not quite perfect, else nothing could ever change. No, it is chaotic, disturbed, even turbulent, like our own seas, our own atmosphere here in the ZMD. And within that turbulence, perhaps even driven by it, small concentrations of energy arise out of random reverberations, reinforcements, and resonances and combine to generate even higher energy concentrations in the field. These concentrations distort the field in their vicinities, increasing its energy density and thus the probability of additional concentrations or strengthening of the initial one(s). By a process of mechanisms similar to those of cellular automata, aggregation, and phase transition these initial concentrations increase their energy level, their stability and their persistence, becoming temporarily stable points or condensations in the field. This undoubtedly occurs on the order of billions of instances in the vastness of the cosmos, but like all similar processes, only a few that achieve stability would be necessary to sustain continued growth.

A two-dimensional diagram illustrates how reinforcement comes about.

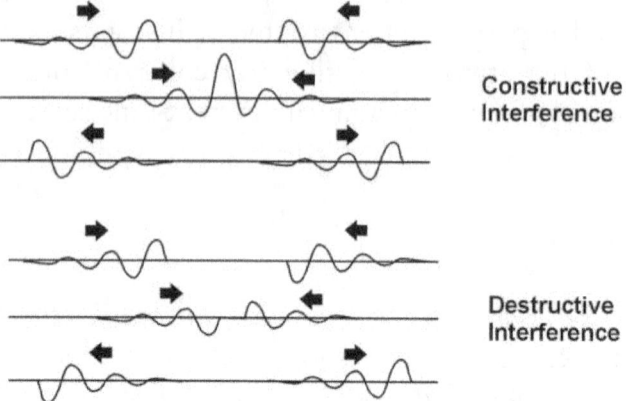

Constructive
Interference

Destructive
Interference

It seems highly likely that the same simple rules or laws of physics might govern this process at all levels of magnitude, tempered, of course, by the wide variations in size, pressure and temperature, in scale all the way from these tiny beginnings to those of massive suns, stars, and galaxies. Like the simple rules of cellular automata, these rules govern local levels of energy and proximity needed to trigger the combination, formation, magnitude, and relative stability of new entities. Heat of transformation levels would govern phase transitions. The intensity of the electromagnetic density of field distortions would either encourage or serve to damp these processes. As in fractal geometry, the same patterns will emerge as one moves from cosmic star-to-star distances down to submicroscopic scales. These rules are probably not digital, that is, not either on or off, 1 or 0. They are not "quantized," that is, they do not divide reality into tiny free-standing increments, based solely on whole numbers and their ratios, as the Pythagoreans favored. But because of the very nature of phase transitions, some parts of the growth process will likely appear step-like. On the other hand, they will, for the most part, describe a universe that is smooth, continuous, describable mathematically only by means of the calculus with its assumptions of infinitesimals.

In this regard, a short recent piece by Amir Alexander in *Scientific American* highlights a possible turning point in our scientific history, a place where we might have first gone wrong:

"Sometime in the 5th century B.C. the Greek philosopher Hippasus of Metapontum, a member of the secretive Pythagorean brotherhood, left his home in southern Italy and boarded a seagoing ship. We do not know why Hippasus was traveling or where he was journeying, but we do know he didn't make it. According to the legend, once the ship was far from shore the poor philosopher was set upon by his fellow Pythagoreans and tossed into the sea. *(Thankfully, philosophers of today seem to have better impulse control! (cs)).*

"The Pythagoreans had good reason to turn on their brother. Following the teachings of their founder, Pythagoras, they fervently believed that everything in the world could be described through whole numbers and their ratios. But Hippasus had proved that the diagonal of a square is incommensurable with the square's side, or, as we would say today, that the square root of 2 (the length of the diagonal relative to the side) is irrational. This means that no matter how many times the side is divided and how many times the diagonal is divided the resulting magnitudes would never be equal.(24)
—Amir Alexander, *Scientific American*, (April 2014)

Hippasus' insight, (much like that other famous one that arose first in 5th century BC Greece and was ignored for hundreds of years, that of the heliocentric solar system), remained virtually unknown until the 17th century when it was reawakened by a number of mathematicians.

"By 1700 Isaac Newton and Gottfried Leibniz had turned this approach into the powerful algorithm we know as "the calculus," ca-

pable of being applied to anything from the motion of the planets to the vibrations of a string and the flight of cannonballs."
—Alexander, op. cit.

In describing the calculus, Alexander favors a term from common usage that I have generally eschewed in this work, that is, the word "infinity" in the form "infinitesimal." He means, of course, *un-measurably small*, to describe the conceptual units that make up lines (dots) and planes (lines), but by making these units un-measurable, he avoids falling into a *Zeno's Paradox* kind of word game. The "smoothness" of the calculus was important in giving us the ability to quantify and predict the only "smooth" things that the ancients could experience. Benoit Mandelbrot points out that "To early man Nature provided just a few smooth shapes.: the path of a stone falling straight down, the full moon or sun hidden by a light haze, small lakes unperturbed by current or wind." Now we work hard to create smoothness, polished surfaces of pistons or tabletops, the Platonic ideal of perfect form and smoothness. But raw nature remains rough.

Today we use measurable "quanta" in many of our devices, even this computer screen on which these words appear as I type. The screen itself is made up of millions of dots (pixels) that are alternately illuminated or damped by instructions in the machine. This pattern is generally smaller than the resolving power of the human eye so that lines drawn on the screen appear continuous, letters totally connected and smooth in their contours, shades and colors smooth in their gradations. By these devices, an imagined smoothness in the real world is imitated on the "digital" screen. A similar example is the motion picture imitation of motion. A series of distinct images are shown in such rapid succession that the finitely limited ability of the brain to process the information produces a credible imitation of smooth motion. It requires about 150-

200 milliseconds for the human perceptual system to process an image and deliver to consciousness as a completed "frame" of a movie. So movie frames (still images) are projected at a least 10 times that rapidly, at 24 or 25 frames per second., so they appear to us as smooth as the reality that has been transferred into a digital, quantized form. This same illusion of smooth motion is apparent in the movement of the arrow-like cursor that tracks across the screen as you move the computer's mouse.

So, the difficult concept to grasp is that a "fine-grain" structure of the universe, like uncutable "atoms", is non-existent. At the scale of the field the universe *is* smooth and connected. At the first stepwise level of phase transmissions that changes, and the distinctions we make lie in their varying magnitudes, the frequencies of their vibrations, and in the complexities of their combinations.

So here are the real "first-beginnings". They are those tiny blinks of light resulting from the random reverberation, reinforcement and resonance of waves of energy in the turbulent sea of the ether, that both contains us and permeates us. As we said in the introduction, they are like what one sees against one's eyelids when the bed-lamp is first turned off. And those are like the lights reported by our lunar astronauts on their return from a moon mission. Lights they "observed" through the ports of their spacecraft while it was outside of earth's magnetic field but which did not appear in their attempts to photograph them. Those, of course, were triggered internally in their perception systems, by firings of brain cells that then faded, caused, in the astronauts' case, by 'space noise" (cosmic rays or whatever you wish to call them), striking their visual cortices while their brains were unshielded by the earth's magnetosphere.

imagine darkness

1.4 *The Alternative Models*

HOW THEN DO WE GET FROM THESE TINY BEGINNINGS to the massive, almost un-measurable construct that we call the universe? Did it happen suddenly, in a cataclysmic explosion like what is called the big bang, arising from nothing and expanding from nothing to something in a microsecond to a vast extent and then beginning its accelerating growth of over 13.7 billion years till today? This is, of course, what is now called "the standard model" of cosmology, generally accepted as the base of knowledge among physicists, astronomers, and cosmologists. Or was it by one of the other modern creation myths like those in what Wikipedia calls *Non-Standard Cosmologies*?

Before we put or own model out there for discussion, it might be useful to discuss a few of these others, to give an idea of the range of speculation that exists and how each of them might explain both our physical and astronomical observations.

In the first place, it is difficult to look at the historical record and separate "creation myths" from "grand unified theories" when talking about the universe. The early models all tied the concepts together. The Gods created, using just a few elements, say earth, air, fire, and water, and for some the invisible substrate, the ether. For others, particularly the "one god" believers, everything just flowed from his (or her) mind, and is just "here." Many of these are still with us. Most of these myths and theories, some more plausible than others, have been have described at length, so here we'll just deal with that question by skipping over a few hundred years and start with the modern models, say, that of Newton.

1. Newton's Universe

Isaac Newton's universe was made up of physical objects, events, and phenomena that existed in a fixed, absolute space for what could be described as fixed, absolute duration. When queried about the thing called "time" he described it as possibly meaning just that, "Time might also be considered as a duration," he said.

Those Newtonian durations, necessary for describing motion and calculating velocity and acceleration, could be seen as made up of periodic or successive intervals, but those hours, minutes, and seconds were arbitrarily devised as just measurements, fractions of commonly observable units such as the passage of the seasons, the rotations of the earth around its axis and around its sun. He did not see space as having physical substance or properties, either, but did think of it as filled with a substance he dubbed the "lumeniferous ether," to explain the propagation of light, heat, gravity, and the like. While truly at most times a pure scientist, he did, however, give in to mystical instincts in writing about things like astrology and the transmutation of elements, for example, and seemed content with not proposing his own creation myth. He satisfied himself with describing things as they are, their effects on each other, their physical relationships, and the mathematical means for calculating those physical values. In other words, he gave us an amazingly complete model of how the universe worked, objects at rest and in motion, the marvelous composition of light in all its colorful parts, the motions of the planets and their moons, but nothing of the universe's origins.

Looking back, his work on gravity was perhaps his greatest contribution. In his studies of the orbits of the planets, Newton showed us most clearly the workings of this mysterious force, that doesn't just make everything fall "down," as Galileo showed us, but when combined with the planets' orbital motions, holds the solar system together. And he taught us how to measure and predict it and its effects. We have worked confidently with Newton's universe for 300+ years now. As Elizabeth Rosner said in her novel, *The Speed of Light*,;

> *Physics floated in my head.*
> *Tendencies to exist. Tendencies to occur.*
> *In the zone of middle dimensions, in the realm of our*
> *daily experience, Newtonian physics is still a useful*
> *theory. We are solid material bodies occupying empty*
> *space.*
> *It comforts us to believe this. (1)*

2. Einstein's Universe

In the early 20th century, Albert Einstein, bless his heart, disrupted that Newtonian sense of comfort. His predecessors in the nineteenth century had expanded Newton's mathematics, came to a measure of understanding of electricity and magnetism, though not of gravity, and actually changed Newton's concept of light from his corpuscular notion back to a wavelike concept. Just as Newton had explained his successes as having resulted from his "standing on the shoulders of giants," Einstein was able to build on the work of some magnificent minds. Faraday, Hertz, Poincare, Minkowski, to name a few. Besides his work on special relativity, tidying up and expanding Galileo's concepts, Einstein adopted Planck's quanta, imagining the photon as a singular unit of energy, explained its relationship to other particles like the electron in his work on the photoelectric effect, showed that matter and energy were inextricably joined by $E=mc^2$, and stirred up controversy with his views on simultaneity and observations between differing frames of reference, and came up with a new, non-Newtonian theory of gravity.

In Einstein's General Theory of Relativity, space and time were no longer absolutes and they, like matter and energy, were tied together in what he named a spacetime continuum, a four dimensional entity that could be curved, warped, interrupted, distorted. Suddenly these two entities, without previously having any actual observable physical characteristics, became the key players in a new concept of exactly what the world was, in all it's and the universe's parts.

An aside: at first this new model did not have an ether. In some ways, Einstein's relationship with the concept of an ether was like that he apparently had with his wives. Attraction, acceptance, denial, then re-embrace. In Special relativity, in 1905, he felt that an ether was unnecessary. After all, Michelson and Morley had effectively erased it from physics in 1887. But after the apparent success of the General Theory in 1916, he rethought that choice. Later in 1916, as we reported earlier, he concluded that spacetime by itself was not suffi-ciently physical to support all of the results that GR described or predicted. "General relativity without an ether is unthinka-ble," he said, in a letter to Hendrik Lorentz.

Still, even with an ether, Einstein's universe was a radical departure from that of Newton. In Einstein's universe, gravity was not an attractive force between masses as in Newton's. Rather the apparent pull of one planet or sun to another was the result of a warping, a curvature, of the spacetime continu-um around masses, that tended to draw them together, as balls placed on an elastic sheet distorts that surface so that objects dropped on it or passing each other tend to run to-gether. (I remember wondering, however, what force pulled down on those little ball bearings so that the sheet did get dis-torted.) But Newton's math still worked to explain the meas-ure of the apparent attractions and to describe planetary or-bits. General relativity also explained other phenomena in a new way. Einstein said that his interest in these notions arose from his teenage wonderings about what it might be like to travel, say in a train, at the speed of light so as to observe how that might work. His conclusions in GR were that as one ap-proached that velocity, objects would be physical diminished in size in the direction of travel and that time also would slow down accordingly. But for Einstein the key difference with Newton's universe was that space and time (Raumzeit) were not fixed, absolute, but were rather active participants in the forces and movements of the world. And even though he rec-ognized the need for an ether as a physical entity to give those characteristics to spacetime, he minimized its role as he con-tinued to seek a way to combine his vision, relativity, to the other recognized forces seen to govern the behavior of matter at what was beginning to be called the "quantum" level.

3. *The Quantum Universe*

Almost in parallel with Einstein's work, a new vision of how the world works was emerging from his and Max Planck's use of the term quantum. Seen initially by Planck as the smallest unit of measurement of energy, then taken by Einstein to represent the energy of a photon, the term quantum began to take on physical characteristics of its own. As with many otherwise purely mathematical terms, it came to have, somehow, by simple usage, and by substitution for reality, or perhaps just as a convenient shorthand among those who spoke mathematics as fluently as any of their other languages, an entity with first virtual reality, then real physical characteristics. Amazingly it is a word that has come to represent a universe of vast extent as witness that we now have quantum leaps, quantum gravity, quantum computers, and all of the varieties of physical theories one can imagine, among them quantum mechanics, quantum electrodynamics, quantum field theory, quantum chromodynamics, quantum levitation, on and on ad infinitum (if you'll pardon the expression). If you wish something to be seen as right *with it*, esoteric, far out, mysterious, it seems that you have only to add quantum to its name.

At any rate, quantum theory development became a serious business, to the extent that we now have many calling it the most successful theory ever devised, with uncountable practical applications in electronics technology, engineering and medical devices, in communication and control systems and the like. I am convinced that many of these very workable applications are based on misconstruals of what is actually true, just as the theories are based on major misattributions of the observations that led us to this point. The fact is, even though their applications work, quantum theories are based on multiple logical paradoxes, contradictions, almost mystical assumptions, even deliberate misreadings of other accepted theories, that students of modern physics, by their own accounts are often advised, "Don't concern yourself with the theories, just calculate!"

These issues deserve a complete book of their own, and in fact thousands have been written. This is not one of them. What we want to show here is how this particular vision fits in the panoply of models of the universe and whether it can continue to be taken seriously or needs to be replaced. Quantum theories undertake to describe how, at the smallest scale, there exist minute entities called "fundamental" particles, how these relate to one another, what are the forces that bind them together or force them apart and how this complex set of mechanisms work to create the large, complex universe we inhabit. It has its roots in particle physics which itself arose from several hundred years of reductionist science, working down from the identification of chemical elements, their constituent atoms, the discovery of radioactivity, and electromagnetism, all linked together in modern chemistry and physics.

Quantum theory, in its modern interpretation, identifies matter (sometimes in its energy phase, see $E=mc^2$) as made up of what now number about sixty-one fundamental particles. Its physical principles are inextricably tied to its mathematical expressions, since none of the sixty-one fundamentals have never been observed experimentally, only posited mathematically as necessary for some predicted or observed experimental outcome. In general this structure is as follows. Atoms consist of combinations of two or sometime 3 main particles. An atom's core, or nucleus contains one or more protons, and in all but one or two elements, one or more particles called neutrons. Protons are said to carry a positive electrical charge and neutrons, per their name, carry none, hence they are considered neutral. Balancing the positive charge of the proton, each atom also carries an equal number of electrons, each carrying a negative electrical charge. When first diagrammed by Niels Bohr, the atom was thought o as a kind of planetary system, of electrons orbiting a core (nucleus) of protons and neutrons in specific orbits. Here is a diagram of some of the smaller ones.

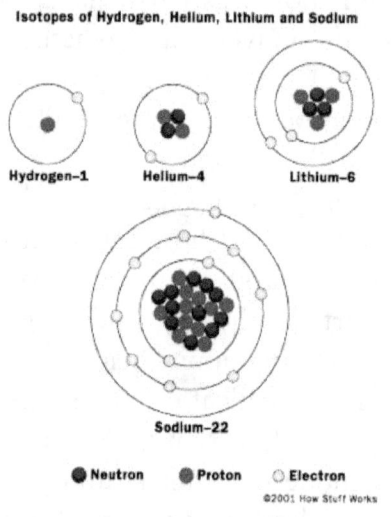

Isotopes of Hydrogen, Helium, Lithium and Sodium

Hydrogen–1   Helium–4   Lithium–6

Sodium–22

● Neutron   ● Proton   ○ Electron

©2001 How Stuff Works

Each of the atoms has an identifying number. That number designates the number of protons in its nucleus and correspondingly, the number of electrons in its orbiting surround. Bohr hypothesized that electrons could move from one orbit to another, but that these levels were fixed, that their could be no gradual movement, only a stepwise one, one or the other, and each orbit was associated with a specific energy level, or quantum. This was the first meaning of what became known as a quantum leap, a discontinuity, not a smooth transition. A second number is customarily associated with each element, its atomic weight. The atomic weight of an atom is approximately equal to the total number of protons and neutrons in the atom's nucleus. The weight of electrons in an atom is for the most part considered insignificant since an electron's weight (or mass) is only about 1/1800th of that of a proton or neutron.

Atoms were, of course, originally considered stable (un-cutable?) but anomalies were soon discovered. The discovery of radioactivity by the Curies, unstable isotopes in chemical experiments, ion transfer, that is, the trading of electrons between different elements, showed that perhaps the atom was not the fundamental tiniest piece of creation, that its constituent parts might be isolated, moved, transferred.

The story of how those constituent parts were discovered, imagined, calculated, invented, predicted, etc., though never actually seen, is one of the most remarkable stories ever told, far more detailed and yes, controversial, than any other creation myth in history. A few examples. It turns out that protons and neutrons themselves are not uncutable. Each is made up of three (exactly) submicroscopic particles called quarks. Now there are different kinds of quarks, necessary because protons and neutrons re different. The same size and weight, but carrying a different charge. So the three in the proton consist of two of the "up" variety and one of the "down" variety" Here is how a proton is described in Wikipedia.

The **proton** is a subatomic particle with the symbol p or p^+ and a positive electric charge of 1 elementary charge. One or more protons are present in the nucleus of each atom. Protons and neutrons are collectively referred to as "nucleons". The number of protons in the nucleus of an atom is referred to as its atomic number. Since each element has a unique number of protons, each element has its own unique atomic number. The name *proton* was given to the hydrogen nucleus by Ernest Rutherford in 1920, because in previous years he had discovered that the hydrogen nucleus (known to be the lightest nucleus) could be extracted from the nuclei of nitrogen by collision, and was thus a candidate to be a fundamental particle and building block of nitrogen, and all other heavier atomic nuclei.

In the modern Standard Model of particle physics, the proton is a hadron, and like the neutron, the other nucleon (particle present in atomic nuclei), is composed of three quarks. Prior to that model becoming a consensus in the physics community, the proton was considered a fundamental particle. In the modern view, a proton is composed of three valence quarks: two up quarks and one down quark. The rest masses of the quarks are thought to contribute only about 1% of the proton's mass. The remainder of the proton mass is due to the kinetic energy of the quarks and to the energy of the gluon fields that bind the quarks together. (2)

The quark structure of the proton. (The color assignment of individual quarks is not important, only that all three colors are present.)

The neutron, on the other hand, consists of one "up" and two "down" quarks. The reason that no one has seen a quark is that they re not known to exist independently of protons and neutrons, Again, from Wikipedia:

"Neutrons bind with protons and one another in the nucleus via the nuclear force, effectively stabilizing it. The number of neutrons in the nucleus of an atom is referred to as its neutron number, which reveals the specific isotope of that atom. For example, the abundant carbon-12 isotope has 6 protons and 6 neutrons, whereas the rare radioactive carbon-14 isotope also has 6 protons but, instead, 8 neutrons. Elements may be found in nature as only one isotope or with as many as 10 isotopes (Manganese and Tin, respectively).

While the bound neutrons in nuclei can be stable (depending on the nuclide), free neutrons are unstable; they undergo beta decay with a mean lifetime of just under 15 minutes (881.5 ± 1.5 s). Free neutrons are produced in nuclear fission and fusion. Dedicated neutron sources like neutron generators, research reactors and spallation sources produce free neutrons for use in irradiation and in neutron scattering experiments." (3)

This is just a taste of the complexity of modern particle physics/quantum physics. In the early part of the 20th century, protons and neutrons were considered fundamental. then other particles were identified until in the 1960's the number of fundamentals rose to the sixty-plus of today, not counting their antiparticle complements. The prolifery of quantum theories noted previously purports to explain how these are bound together, what qualities or conditions adhere to them in their various states of existence (or non-existence, or virtual existence), as well as the limitations on our ability to detect them or their characteristics such as energy levels, momentum, location at any instant, superposition (that is, existence of two or more at the same location at the same time), and whether they should be considered as wave-like or particle-like. Quantum theories have been criticized over their almost 100 years of existence for many things: their inherent inconsistencies and contradictions, their "incompleteness", most notably by Albert Einstein, and their almost mystical insistence on what "cannot be known."

We will be touching on many of these aspects in the following pages as we show how they fall short of fully explaining the world as we know it, or in some cases actually prevent us from seeing the true reality of the world. Of course any "cosmology" or theory of the probable creation and existence of the universe always leads back the question of how and from where did these fundamentals arise, be they atoms or the even smaller particles of which we assume they consist. The closest thing we have to a "quantum" cosmology is of course the theory of the big bang, which in its latest incarnation insists that everything was created out of nothing in the greatest explosion ever known.

4. *Speculative Cosmologies*

Of the universes described above, Newton's and those of the thinkers following him up until the 20th century have become known as the "classical" theories. These have included all of the work of physicists like Faraday, Hertz, Lorentz in electricity and magnetism, astronomy, mechanics and the like. Beginning with Einstein and following on with Bohr, Schrodinger, Heisenberg and many others, we saw the development of what are now known as the "Standard Models" Modern particle/quantum theories at the smallest scale of physics, Einstein's relativity theories and the big bang in particular, at the largest scale of astronomy and cosmology, are now generally accepted as the de facto answers to almost all of the questions about the universe and how it works, with one gigantic problem. All believe that everything is held together or apart by four fundamental forces, a weak force at the sub micro level, about equivalent to the electromagnetic force of magnetism and electricity, a strong force that bind the nuclei of atoms together, and gravity, which among other effects, holds our feet to the ground. The big problem is that relativity, which explains(?) gravity, doesn't explain the other forces, and quantum theory, which tells us how atoms and their parts work, doesn't explain gravity. Einstein worked on this problem for most of the last part of his life and many others have devoted millions of hours and thousand of pages in the effort, thus far to no avail. But even without this serious gap between the theories, relativity does not deal adequately with the world at its smallest level, and none of the quantum theories come near to being complete and free of gaps and contradictions. Some of the serious detractors have tried the Ptolemy approach, of tinkering at the edges, adding little epicycles to smooth out the rough places, others have taken great leaps, mostly, unfortunately, into the great unknown. Expansion has been added to the big bang. We will return to both of those ideas later. For now let's look at the special section on Wikipedia, called "Non-standard Cosmologies."

A **non-standard cosmology** is any physical cosmological model of the universe that has been, or still is, proposed as an **alternative to the Big Bang** model of standard physical cosmology. In the history of cosmology, various scientists and researchers have disputed parts or all of the Big Bang due to a rejection or addition of fundamental assumptions needed to develop a theoretical model of the universe. From the 1940s to the 1960s, the astrophysical community was equally divided between supporters of the Big Bang theory and supporters of a rival steady state universe. It was not until advances in observational cosmology in the late 1960s that the Big Bang would eventually become the dominant theory, and today there are few active researchers who dispute it.

The term *non-standard* is applied to any cosmological theory that does not conform to the scientific consensus, but is not used in describing alternative models where no consensus has been reached, and is also used to describe theories that accept a "big bang" occurred but differ as to the detailed physics of the origin and evolution of the universe. Because the term depends on the prevailing consensus, the meaning of the term changes over time. For example, hot dark matter would not have been considered non-standard in 1990, but would be in 2010. Conversely a non-zero cosmological constant resulting in an accelerating universe would have been considered non-standard in 1990, but is part of the standard cosmology in 2010." (4)

As can be seen from this introduction, the main test for acceptance is scientific consensus, but the primary reason for being called non-standard is simply definitional, that is, they are not the big bang. The fundamental assumptions necessary for any alternative cosmology as follows:

1. the universality of physical laws – that the laws of physics don't change from one place and time to another,
2, the cosmological principle – that the universe is roughly homogeneous and isotropic in space though not necessarily in time, and
3. the Copernican principle – that we are not observing the universe from a preferred locale. (5)

There is a fourth, not listed in the Wikipedia article: we cannot know what preceded the big bang, since time and space as we know them would have had their first beginning at that instant.

These assumptions when applied to the Einstein field equations naturally result in a universe which has the following features:
1, an expansion of the universe,
2. a universe emerging from a hot, dense state at a finite time in the past,
3. the lightest elements were created in the first moments that time existed as we know it, and
4. a cosmic microwave background pervading the entire universe should exist, which is a record of a phase transition that occurred when the atoms of the universe first formed.

These features were derived by numerous individuals over a period of years; indeed it was not until the middle of the twentieth century that accurate predictions of the last feature and observations confirming its existence were made.

It is not clear how these last four conditions are derived from the field equations, particularly as regards numbers 2, 3, and 4. Let's look carefully at the Standard Model. A brief description from a popular web site goes like this:

The big bang
Our universe is thought to have begun as an infinitesimally small, infinitely hot, infinitely dense, something - a singularity. Where did it come from? We don't know. Why did it appear? We don't know. After its initial appearance, it apparently inflated (the "Big Bang"), expanded and cooled, going from very, very small and very, very hot, to the size and temperature of our current universe. It continues to expand and cool to this day and we are inside of it: incredible creatures living on a unique planet, circling a beautiful star clustered together with several hundred billion other stars in a galaxy soaring through the cosmos, all of which is inside of an expanding universe that began as an infinitesimal singularity which appeared out of nowhere for reasons unknown. This is the Big Bang theory. (6)

If read carefully, anomalies arise almost instantly. The big bang model depends on several important constructions, not the least of which is its dependence on "infinities."

First we discover that the universe seems to be expanding, and at an increasing rate. We "know" this because Edwin Hubble saw that the light from distant objects in the universe, the galaxies at the most remote locations he could see exhibited a characteristic "red shift" in their visible spectrums. The red shift was directly equated with their velocity related to us, their earthbound observers. In the same way that the sound from a moving train varies as it speeds first toward and then away from us, its pitch rising as it approaches, then falling as it recedes, the light from a moving source shows the same wave-shortening effect if moving towards us, a shift toward the blue end of the spectrum, then lengthens as it moves away, the so-called red shift. Thorough many observations and measurements Hubble was able to show that the magnitude of the red shift seemed to tell us how rapidly each of those distant objects was receding. And if these objects were clearly moving away from us, then there must have been a time when they were much closer, perhaps even a starting point. When might that have been?

Second, for the universe to have reached the level of mass and energy (remember that these are equivalent, per Einstein's Special Relativity) that it appears to have today, it must have begun from a tremendously hot point at its beginning, and if we can look back far enough, that is, at the objects farthest away from us, whose light has taken perhaps billions of light years to reach us, we may find out how old this universe actually is.

Third, we can tell from observations with our sensitive instruments and techniques the relative abundance of the various chemical elements that exist in the universe, and lo, those are hydrogen, then helium, and lithium. We are right on.

Fourth, let's find a way to detect whether there exists an echo of that original explosion, and that turns out to be easy, two engineers from Bell Labs have found it for us, in the microwave range of frequencies and at about 2.7 degrees Kelvin.

But there are problems.

First, as we look at the skies there seem to be some bright objects out there that, by all the rules we've laid out seem to be older than the age we've calculated for the whole universe, some 13,82 billion years. And also, there are objects called quasars (quasi-stellar objects) that are supposed to be the

originators of galaxies, hence older than the galaxies surrounding them, but that have much smaller red shifts, meaning younger, mixed right in with those much older stars.
Maybe the red shifts come from some other source. What has
been forgotten in this discussion is that the man who gave his
name to this effect was, in fact truly skeptical that red shifts
actually correlated with expansion. Hubble was deeply engaged in a study that would have confirmed or disproved that
thesis at the time of his death, and had expressed his doubts
on numerous occasions. But those studies died with him and
his successors ignored those doubts.

Second, what was in that hot point that triggered that first
mother of all explosions? Could this all have come from nothing? Well, it isn't any more paradoxical than some of the other quantum theory stuff so let's just call it a singularity and
leave it at that.

Third, when we trace the expansion trail back, our lines
don't meet at a point but back to a larger area right after the
beginning. But Alan Guth has an idea. What if there was in
fact a big inflation right at the beginning when all those light
elements spread out and then the expansion began but from a
fatter bubble? That makes the math work and no one can
prove it didn't happen that way.

Fourth, we've got that CMBR from Penzias and Wilson,
but people were finding something just like it coming from the
stars, coming from reflected light in the sky, even coming
from the earth itself, and at the same frequencies and temperatures, as long as a hundred years ago. How can we prove it's
the echo of the big bang? Well, we can't, but again, it makes
most of the math work, so we'll use it anyway.

The trouble is, people kept coming up with items like these
until there were too many of them to ignore, and besides that
there were new models coming along that were more con-

sistent and answered more of the questions. More and more the big bang has come into question. Appendix 1 goes into these in more detail

The suspicion is that these had to await some more observed or imagined evidence, since their prediction by the field equations directly strains plausibility. For example, the so-called Cosmic Microwave Background Radiation was supposedly detected serendipitously at a later date, was originally attributed to unknown causes until someone said, "Oh, that's in a range that would have been predicted from a big bang!" and has since been seen as actual confirmation that there was a big bang. Which came first? As we will show, an alternative explanation exists.

So, besides the standard model, the big bang, several alternative hypotheses of the origin of the universe are listed, grouped according to the manner in which they contradict the big bang, the de facto standard.

The first group contains alternative metric cosmologies, which includes a range of models, both Newtonian gravity-based through modified relativity-based models from Lorentz, De Sitter, Mach, Godel and others. These are generally based on known geometries and vary according to how they define the forces, gravity, in particular, that hold the universe together, its orbits and general effects on its different parts. They nearly all share the assumption of expansion, supposedly proven by Hubble's constant.

The second group includes the "steady state" models, primarily initiated by Fred Hoyle. From Wikipedia:

"The Steady State theory was proposed in 1948 by Fred Hoyle, Thomas Gold, Hermann Bondi and others as an alternative to the Big Bang theory that modified the homogeneity assumption of the cosmological principle to reflect a homogeneity in time as well as in space. This "perfect cosmological principle" as it would come to be called predicted a universe that expanded but did not change its density. In order to accomplish this, steady state cosmology had to posit a "matter-creation field" (the so-called C-field) that would insert matter into the universe in order to maintain a constant density.

The idea was almost immediately attacked by proponents of the Big Bang who described the C-field as contradictory to a consistent understanding of physics. Hoyle, one of the most vocal proponents of the steady state model, and a committed materialist, believed that the competing, older model was forced as it violated fundamental philosophical principles regarding the infinite nature of existence. Hoyle explicitly warned that the Big Bang was being promoted as a first cause dogma in line with Western theology rather than science." (7)

What we have here is, of course, a fundamental disagreement about origins. The big bang (a name actually coined by Hoyle in a disparaging comment in his dissent from the theory, but one that unfortunately stuck around) requires an assumption that "something came from nothing," not far from God saying, "Let there be light." Hoyle's steady-state universe needed a "c-field," also an act of faith. Neither could point to a precedent or evidence of something already in existence that could have been the generator. Assumptions begin to pile up. Again from Wikipedia:

"The debate between the Big Bang and the Steady State models would happen for 15 years with camps roughly evenly divided until the discovery of the cosmic microwave background radiation. This radiation is a natural feature of the Big Bang model which demands a "time of last scattering" where photons decouple with baryonic matter. " (8)

Now note the assumptions backing the big bang: "background radiation," a natural feature of the big bang model , " photons," and "baryons," essential assumptions of quantum/particle theory. Again no alternative sources are sought.

Finally, there is a group of alternate cosmologies based on skepticism concerning all or parts of some of these prior assumptions. Three of these emerge from skepticism concerning the expanding universe idea. Hubble derived his redshift theory from the wave theory of electrodynamics, using the analogy of sound waves in the atmosphere whose pitch is lowered as the source moves rapidly away. Light coming from a source moving away from us at a substantial velocity should exhibit a shift toward the red end of the spectrum. And lo, observations of stars and galaxies thought to be at great distance away showed just such a shift. And the differences could be measured and the relative velocities calculated. More distant points of light were clearly moving away faster. Expansion was confirmed. The skeptics questioned 1, whether light coming from such distances might exhibit a redshift due to energy losses from the long passage though local gravities, electromagnetic fields, cosmic dust, and the like. This was dubbed the "tired light" hypothesis. Others, notably Halton Arp, observed quasars in the midst of distant galaxies that Hubble's redshift theory would have placed much closer than their surrounding stars. Could this mean that redshift was not caused only by receding velocities but might be inherent in younger stars, changing as they aged. This debate is still going on. Maybe the universe is not expanding.

Other theories based on physical observations posited a plasma universe, filled with an energy plasma. This theory shares with the steady-state theories the assumption that the universe and the matter it contains has always existed, whatever that means. So we shouldn't be concerned about an origin, there never was one.

Finally, though not included in this group are another set of cosmologies that have grown in more modern times, perhaps out of frustration that the standard models can't seem to be reconciled. These include string theory, now three or more decades old without yet having produced a single testable assumption, described by the physicist Lawrence Krauss as resulting from throwing a dart against a blank wall and then drawing a target around the point of impact. There are multiverse theories suggesting that ours may be only one of many universes and that some of the others may have totally different physical laws; one that suggests that our universe may have as many as 32 additional dimensions explaining why we can't figure out all of the details because we can't access those other (astral?) planes.

All of these have gained the distinction of being classed by the writer and science critic Jim Baggott as "fairy-tale physics," based on multiple levels of unsupportable assumptions, and as a result, not worthy of any support. The unfortunate part of this is that some of them, most notably string theory, have somehow grabbed most of the educational and government grant supply with the result that other research efforts have found themselves starved and weakened. For a fuller account of this phenomenon, I would recommend 4 books, Peter Woit's *Not Even Wrong*, Jim Baggots's *Farewell to Reality*, Alexander Unzicker's *Bankrupting Physics*, and for a somewhat gentler view, Lee Smolin's *The Trouble with Physics*.

Other alternative models have been proposed, some with similar roots as the simple universe. A particular school that comes somewhat close is that of Gabriel LaFreniere (1942-2912) whose thesis is that "Matter is Made of Waves." Unfortunately his and his followers' work stops very short of being complete. For LaFreniere the fundamental particle is the electron, which arises out of the aether as a wave phenomenon, but the aether is an unspecified substance, only that it must be there for an electron to be made from it. The work is long on details and the mathematics of wave mechanics, but short on substance, so it falls short of being considered physics. (see note)

So is there a place for a new cosmology, a new physics, even a new unified theory? Richard Feynman perhaps put his finger on it when he suggested that the answer probably lies in a whole new direction, that tinkering with the old theories may be reaching a Ptolemaic epicycles stage. Let's assume that any cosmology must meet the same principles asserted earlier for the big bang, that is:

1. the universality of physical laws – that the laws of physics don't change from one place and time to another. Ideally these laws should apply at all scales.
2, the cosmological principle – that the universe is roughly homogeneous and isotropic in space though not necessarily in time.

The first two are generally combined, as in Astronomer William Keel's summation:

"The cosmological principle is usually stated formally as 'Viewed on a sufficiently large scale, the properties of the Universe are the same for all observers.' This amounts to the strongly philosophical statement that the part of the Universe which we can see is a fair sample, and that the same physical laws apply throughout. In essence, this in a sense says that the Universe is knowable and is playing fair with scientists."

A principal difficulty with number two is that there is truly no observable confirmation of this notion. In cosmology, in particular, there are large, fundamental gaps and voids and large concentrated clusters. The cosmological principal works for the math, but not for reality. Left to its own devices, nature is uneven, and rough. We will be demonstrating this subsequently.

3. the Copernican principle – that we are not observing the universe from a preferred locale.
4. we cannot know what preceded the beginning, since time and space as we know them did not exist prior to that instant.

The third, the Copernican principle, is important because it effectively gets rid of a number of things, religion; the anthropic principle, that is, that the universe developed in this way just so there would be something for we humans to observe and wonder about; and the belief, in some quarters, that where we are is in effect, the actual center of the universe. The fourth principal, added by this author, is a major part of the big bang theories, less so in steady-state theories, with their assumption that everything has always been there, and will perhaps turn out to be subject to question as we go along. Key to this answer will be how we end up defining time. Here, I would like to insert a fifth principle: that of logical plausibility. An acceptable cosmology, a model of what we are and how we got here, so to speak, should meet the test of what in Italian is known as *buon senso,* literally translatable as "good sense," but meaning logical, obvious, clear, believable without stretching the bounds of imagination.

To that end, that one of our principal goals in this effort is the achievement of *buon senso,* a further statement is required. To the critical eye and ear, a principal failing of every current theory or model of the cosmos, every cosmology or physical theory currently accepted as a "standard model," is their dependence on one or more patently unverifiable, untestable assumptions. Until Einstein recanted his denial of the existence of an ether, his theories were based on assumptions that two fundamental foundations of those theories, space and time, had real physical attributes. In most current formulations and references to relativity that untenable assumption is retained. Quantum theories, all of them, have their own untestable assumptions, wave-particle duality, wavefunction collapse, the role of the observer in affecting the outcome of observations, to name a few. The big bang is based on possibly the greatest of these untestable assumptions, that everything came from nothing. The model of the universe we are building here, on the other hand, the descriptions and explanations of the operations and mechanisms taking place in the invisible world, both that part that is too small for us to see and that part that is too far from us to directly examine, is based entirely on the congruency of the laws of physics in those realms with those operational in the world we <u>can</u> see and examine *here*, in the zone of middle dimensions. If I have a fundamental goal, it is the same as that of Thomas Kuhn who said in the Preface to *"The Structure of Scientific Revolutions"* ". . . my most fundamental objective is to urge a change in the perception and evaluation of familiar data." The premise here is that there is a fractal-like correspondence at every scale of existence, and with that as a starting point, the possibility exists that the rules, the real laws of physics, may turn out to be relatively few and very simple.

So, keeping these principles in mind, let's talk about beginnings.

Part 2. The Assemblage—The Universe as a Self-organized System

"Most generally, the harmonies that Kepler saw in the planets' motions have largely been discredited, yet it remains very broadly the case that the starting point of every science is to identify harmonies in the raw mess of evidence."

—Benoit Mandelbrot.

imagine darkness

2.1 How Did It Start?

WE BEGAN THIS DISCUSSION with the premise that there exists a physical reality outside of our heads and that reality exists as the set of all observable entities we describe as objects, events, and phenomena, those we apprehend only by their effects, like electrical and magnetic fields, those we clearly observe in the flesh, so to speak, that exist and inhabit our perceived world, on up to those of an astronomical scale, that we perceive only at great distances.

And you will recall that we earlier arrived at *Axiom 1*, a statement that something actually existed before there was a universe as we know it. That there was and still is today, a fixed, essentially limitless electromagnetic field out of which all the reality we perceive arose. We showed that such a field can be seen to exist, not as the multiple "quantum" fields assumed to be singularly linked to each of many kinds of particles, interacting with them and each other, but rather as a single fundamental field, and what we observe as differences are in fact simply complex distortions in that field.

We then showed in **Axiom 2** that it was highly probable that the assumed fundamental building blocks of the standard model, those things called particles, were not necessarily the only way of thinking about the structure of things. We saw how a uniquely observable feature of the classical electromagnetic field, its wave-like manifestation, could easily be perceived and mistaken for, the ancient atom, the "first-beginnings" of Democritus, Epicurus, and Lucretius, even in their modern verbal incarnation as quanta, photons, quarks.

Which brings us to Axiom 3

Axiom 3. The region that we call the universe, including all of reality that we perceive around us, is made up in its entirety of these complex coherent organizations of energy, these "condensations of the ether" as Einstein called them. In fact, all identifiable components of reality, all of the objects, events, and phenomena, consist of coherent deformations of the ether, from the smallest entities we can identify, through those that make up what we have called the "zone of middle dimensions," out to the most distant and immense features of the cosmos, the stars, quasars, galaxies, and clusters of the astronomers.

Until the arrival, in the 1960's of what became known as chaos theory, steeped as I was in the second law of thermodynamics and its unbreakable rule that all descends to maximum entropy, the eventual heat death of the universe, I often found myself wondering how any order could possibly arise and persist in the face of this inexorable decline. Where did anything come from? Surely enough time had gone by in the billions of years of history that all would certainly have come to an end by now. But lo, not only were we still here but apparently new stars were being born and matter was being created out of something. Lucretius' dictum that, "nothing can arise from nothing" certainly meets the criterion of logical plausibility (unless you are Lawrence Krauss), so where can it be coming from? Order clearly arises out of chaos but how? Under what conditions?

What chaos theory tells us is that 1) chaos is by definition not uniform. It is characterized by something about which we know a lot, but not enough, and that is the phenomenon of turbulence. When you put the chocolate chips in your cookie dough and stir until they are absolutely evenly distributed, the same distance apart, they are not *randomly* distributed. Both their locations and their distance relationships must be random. The earth's atmosphere is not evenly distributed. Meteorologists have struggled with turbulence since the beginning of their discipline. The oceans stay in one place but they are beset with currents and other disturbances. There is an account (probably apocryphal) of a friend's conversation with Einstein in which the great physicist suggests that if there is a heaven, he would like to ask God about two things, first, "Was I right about relativity?" and then "what about turbulence." The friend replies, "OK, ask about relativity, but don't embarrass God by bringing up turbulence." We recognize that it exists, but we still do not have an adequate calculus to describe it. Turbulence is the classic ill-structured problem. Perturbation Theory can come close but will never give an absolute solution. But, and this we can see demonstrated in multiple situations, out of turbulence, the true chaotic system, order can spontaneously arise.

James Gleick in his book *Chaos*, (1) gives not only an account of how this new science came to be but countless examples of its workings. We mentioned one such earlier in reference to the experiments of Balthasar van der Pol, who found that by increasing the energy level to a vacuum tube system the output, detected through an ordinary telephone handset periodically changed from auditory "noise" to pure, coherent sound at recognizable pitches. Here the causal variable was something we will call *energy density*, a key measurable feature of electromagnetic fields. As Gleick puts it, van der Pol was "one of many scientists who got a glimpse of chaos, but had not the language to express it."

Per Axiom 1, this hyperuniversal field, call it the ether for now, is not just everywhere, but it is everywhere turbulent. And since it is also everywhere pure energy, if any order is to arise in it, that too will be composed of energy. The so-called "condensed matter theory" makes a stab at this but it, like most other aspects of the standard model, attempts, not very successfully, to tie it with the multiple particle assumptions of quantum mechanics.

Chaos theory actually dates back to the 1880's with Poincaré taking up the same complex riddle that had baffled Newton, the three body problem. Simply stated both asked, "How does one calculate or predict the gravitational effects of both the sun and the earth on the orbit of the moon?" Poincaré found that there can be orbits that are nonperiodic, and yet not forever increasing nor approaching a fixed point. Their precession and patterns were not predictable. The problem then was, as James Gleick put it in regard to the discoveries of Balthasar van der Pol, the observations were there but the cross-disciplinary connections and language to describe them had not yet evolved. The kind of computations required were long and tedious and the tools were not yet available. A fascinating parallel in astronomy was the advances that resulted from a lifetime of painstakingly detailed, meticulous observations and records of Tycho Brahe, the sixteenth century Danish astronomer, who never developed a model or theory from his data but laid the foundation for brilliant work by those who followed him. In the case of chaos theory many researchers joined the search, from generally related efforts like Bertalanffy's *General Systems Theory*, (2) along with mathematicians seeking ways to model the new data.

"(Most of) these studies were directly inspired by physics: the three-body problem in the case of Birkhoff, turbulence and astronomical problems in the case of Kolmogorov, and radio engineering in the case of Cartwright and Littlewood.[Although chaotic planetary motion had not been observed, experimentalists had encountered turbulence in fluid motion and nonperiodic oscillation in radio circuits without the benefit of a theory to explain what they were seeing." (3)

The real catalyst for the development of chaos theory and complexity science which grew out of it was the development of the digital computer and methods of computer simulation of natural occurrences, developed primarily to support the nuclear weapons program in World War II. The "father" of what finally became formalized as chaos theory is generally regarded to be Edward Lorenz, who in 1961 was running computer simulations on an early digital machine in attempts to model the origins and development of weather phenomena. After running one simulation, which in those days of primitive computer systems sometimes took days of computer time, Lorenz decided to shorten an experiment by restarting it in the middle of its run. The original simulation had used a six-digit value as one of its parameters, 0.506271. By restarting from a printout of the original results, the parameter value had been rounded off to three digits, to 0.506, a tiny variation not thought to be significant. But when the simulation was completed the outcome was vastly different from the original test. A tiny difference in initial conditions had made an enormous difference in the outcome, a happenstance later somewhat romanticized as "the butterfly effect" where the flap of a butterfly's wings in Brazil could result in a tornado in Oklahoma. It showed that even detailed atmospheric modeling cannot, in general, make precise long-term weather predictions.

A computer simulation is, of course, not an exact reproduction of the real world. First a computer model must be built with the necessary algorithms and equations to approximate, in this case the beginning conditions of the process to be simulated. This requires some guesswork, although researchers tend to not use that language. Attempts are made to predict changes in the early behavior of a system, usually one that proceeds through multiple iterations of a particular process. What is common in the process we are discussing is that the particular algorithms are designed to allow the impact of noise to enter the system, matching if possible the observed presence of noise in the natural system being modeled. The use of computers to run rapid multiple iterations of a repetitive process is the factor that gives simulation its great power. And the fact that noise enters the process randomly is what limits its predictive capability. The scale of events being simulated by computer simulations far exceeds anything possible (or perhaps even imaginable) using traditional paper-and-pencil mathematical modeling. First extensively used in this country in the Manhattan project in World War II to simulate possible nuclear explosion problems, the use of computer simulations rapidly spread into the mathematical modeling of many natural systems in physics (computational physics), astrophysics, chemistry and biology, human systems in economics, psychology, social science, and engineering.

Developments in this new way of seeing the world came rapidly. In 1963 Benoit Mandelbrot began studies that resulted in what he called fractal geometry, showing that certain patterns in nature have the property of scale-independence, that is, the same patterns are repeated at all scales of development. Mandelbrot's book, *The Fractal Geometry of Nature*, (4) became a classic of chaos theory. Biological systems such as the branching of the circulatory and bronchial systems proved to fit a fractal model.

The common thread here is that the definition of chaos theory that evolved was very simple. It described systems, an increasing number of them across many scientific disciplines, where, "Small differences in the initial conditions of a dynamic system result in great changes in outcomes."

1. SOS—Breaking the 2nd Law

"WHY IS THERE SOMETHING RATHER THAN NOTHING?" That is the question that has bothered philosophers as long as philosophy has had a name. Answer that and you have set the stage for thousands of other questions (and maybe answers). One of the toughest barriers to cross, however, is contained in the "truth" we mentioned back in the Preface., in the Laws of Thermodynamics.

And here's the reason why. The *First Law of Thermodynamics* is a statement of the conservation of energy, that in *any closed system*, there is a limit that cannot be exceeded, that while energy and mass might be interchangeable, there is a fixed supply of both in the universe. The *Second Law* is a statement about the direction of that conservation, that it cannot be reversed – and the *Third Law* is a statement about the impossibility of reaching Absolute Zero (0° K).

The British scientist and author C. P. Snow had an excellent way of remembering the three laws:

Law 1. You cannot win. (that is, you cannot get something for nothing, because matter and energy are conserved)

Law 2. You cannot break even. (you cannot return to the same energy state, because there is always an increase in disorder; entropy always increases)

Law 3. You cannot get out of the game. (because absolute zero is unattainable)

The second of these is what's frightening. It's all down hill. Given enough time and without an infusion from outside, the eventual outcome is inevitable, everything stops, this is the heat death of the universe. What the third law says, then, is if that happens we'll never know because we'll be long gone, there'll be no one around to see the thermometer reach that point. So, given that this universe appears to have been around for at least 13 billion years, why is everything still here? How long is long enough? There's an escape hatch in there, however, maybe two or three. The first is that conditional clause, "in any closed system." If the universe is finite, as we suggested earlier, there might be a source of energy that is feeding it from outside. We've already suggested that. The second is that the idea of an "absolute" zero is just true in the mathematics. We know that the math world is really an idealized construct, not a real world. It even sometimes throws out values like "infinity," also not findable in reality.

So what has research in the real world come up with to explain the continued existence of "somethings?" Well, under the general rubric of "chaos theory," a discipline known as Self-organized systems (SOS) has surfaced.

"Self-organization can be defined as the emergence of coherent, global behavior out of the local interactions between components of the system. This emergent organization is characterized by intrinsic autonomy, adaptability to environmental changes, and local awareness of the most important global variables. Most importantly, many SOS appear to be robust with respect to a variety of disturbances and intrusions, as the system is to some degree capable of overcoming or self-repairing damages. While many natural, social and technological examples of SOS exhibiting these characteristics are known, and several mechanisms of self-organization have been analyzed in detail, the design of a SOS remains a fundamental challenge". (5)

In fact the word *design*, as applied to a self-organized system, is a contradiction in terms. Self-organized means "not designed," just occurring naturally

I think it has been established, if not generally recognized for purposes of analysis, that the universe qualifies as a complex, dynamic system. Complexity and complex systems, generally refer to systems of interacting units that display global properties not present at lower levels. These systems may show diverse responses that are often sensitively dependent on both the initial state of the system and nonlinear interactions among its components. From many other sources, particularly from studies in fractal geometry and cellular automata, both to be discussed later, we can see that complex behaviors can emerge even though the system components may be similar and follow simple rules. This process is known and recognized as *self-organization*. Wikipedia introduces it like this:

"**Self-organization** is a process where some form of global order or coordination arises out of the local interactions between the components of an initially disordered system. This process is spontaneous: it is not directed or controlled by any agent or subsystem inside or outside of the system; however, the laws followed

109

by the process and its initial conditions may have been chosen or caused by an agent. It is often triggered by random fluctuations that are amplified by positive feedback. The resulting organization is wholly decentralized or distributed over all the components of the system. As such it is typically very robust and able to survive and self-repair substantial damage or perturbations. In chaos theory it is discussed in terms of islands of predictability in a sea of chaotic unpredictability. (6)

Self-organization occurs in a variety of physical, chemical, biological, social and cognitive systems. Common examples are crystallization, the emergence of convection patterns in a liquid heated from below, chemical oscillators, swarming in groups of animals, and the way neural networks learn to recognize complex patterns." (6)

Hermann Haken, of the Institute for Theoretical Physics of the University of Stuttgart, Germany, puts it this way:

"Self-organization is the spontaneous often seemingly purposeful formation of spatial, temporal, spatiotemporal structures or functions in systems composed of few or many components. In physics, chemistry and biology self-organization occurs in open systems driven away from thermal equilibrium. The process of self-organization can be found in many other fields also, such as economy, sociology, medicine, technology.

Many objects in our surrounding and daily life such as furniture, houses, cars, TV-sets, computers are man made. On the other hand, especially in the animate world, objects grow, acquire their form, and function without being created by humans. The animal kingdom abounds of examples. It is increasingly recognized that even the human brain may be considered as a self-organizing system as well as quite a number of manifestations of human activity, such as in economy and sociology. But processes of self-organization can be found also in the inanimate world: formation of cloud streets, planetary systems, galaxies etc. A fundamental question is: *Are there general principles for self-organization*? In the inanimate world a positive answer could be found for large

110

classes of phenomena. In the animate world so far at least some insights could be gained. In biology (and perhaps other fields) there is a controversy: are there general principles or do we need special rules and mechanisms in each individual case?" (7)

In physics there are many observed examples generally seen as the formation of spatial, temporal or spatiotemporal patterns. Here are a few examples (from the Haken article):

- Lasers: coherent light, self-organization of many atoms
- Nonlinear optics: coherent light, self-focusing, generation of harmonics, coherent Raman and Brillouin scattering, etc.
- Fluid dynamics, gas dynamics: cloud streets, convection instability, Taylor-Couette flow, roll patterns, hexagonal patterns (Bénard cells), weak turbulences, defects, etc.
- Gas discharges: patterns of molecular densities under the impact of electromagnetic fields.
- Plasma physics: density and velocity patterns of partly or fully ionized atoms and electrons in (partly self-organized) electromagnetic fields, instabilities.
- Semi conductors: patterns of electron and hole densities and currents, Gunn-effect, current filaments.
- Astrophysics: formation and structure of planets, stars, galaxies, big bang, voids, etc.
- Meteorology: climatology, cloud formations, cyclones, etc.
- Geophysics / Geodynamics: inner and surface structure of the earth,
- Self-sustained oscillations: can be found in many of the above mentioned fields.
- Radio-engineering and other sources of coherent electromagnet fields: (8)

Although the second law says that the trend toward disorganization to increasing entropy is not reversible, (in C. P. Snow's terms, you cannot win), order *cannot* arise spontaneously from higher entropy in an isolated system, *all of these are known to occur.*

Erich Jantsch laid out the argument in 1980 in his book, *The Self-organizing Universe: scientific and human implications of the emerging paradigm of evolution.*(9) Jantsh's conceptual universe, unfortunately, had its origin in a big bang, not exactly an evolutionary event. However. all of the evidence, all that we so far know about our universe, puts it totally in this class. It is certainly a complex, dynamic construct, and in our model it has arisen from a complex, dynamic, chaotic field of energy. In his classic theory of biological evolution, Charles Darwin postulated no beginning event. For whatever reason, he was more concerned with what occurred afterward, although his theory did assume a progression from simpler to more complex structure as a consequence of adaptive mutations. We are to some extent in the same position. We have a potential starting point, the concentration of energy into coherent, temporarily stable points or regions, depending how intense or distributed these first beginnings of ours arise in the field. We now need to examine some known processes that might lead us to the more complex elements of our current, self-organized entity, the universe. They may even follow some of the same or similar steps we already know from biological evolution.

If we start from the position that our universe is a self-organized system within the cosmos (field, ether), we must also assume that it may have, in whole or in part, self-adjusted multiple times. Some oscillation should be considered normal behavior. In its proto-state, as displayed in the Planck images, it is relatively uniform but contains energetic turbulence, pockets, currents, discontinuities, and what is called 1/f noise (filtered out in the published images).

The existence of a measure of 1/f noise also suggests strongly that this pocket at least, has periodically reached a stage of criticality, where a seemingly minor event may have triggered an avalanche, a collapse, into a new, more persistent, stable configuration. This process, *self-organized criticality* is not time-delimited and not predictable; and temporary stability, on a cosmic scale, can mean billons of years. How this occurs and its implications follows.

2. *SOC—Self-organizing Criticality*

IN THE LAST THIRTY YEARS OR SO, we have begun to understand some of the workings of nature that seem to defy the 2nd law and lead to organization as opposed to chaos. We know a lot about how structural changes can come about in certain materials by phase transitions, in water as we have previously discussed, and in metals, for example. Other discoveries along this line have come from computer simulations that then gave us insight into real workings of natural systems, like cellular automata, network automata, and the development of fractal geometry and its place in the natural world. All of these will turn out to have a place in the issues we are examining. A key part of understanding how self-organization comes about, however, came in the discovery of *self-organizing criticality*.

Chaos theory very quickly grabbed the imaginations of scientists in many fields, from physics to social behavior. This new way of thinking about complex systems seemed to offer a way out of some dead ends of research. It was clear that there were many things that reductionist science could not explain, many dynamic, non-linear processes were being observed and a new way of looking at them was long overdue.

"In December 1977, the New York Academy of Sciences organized the first symposium on Chaos, attended by David Ruelle, Robert May, James A. Yorke (coiner of the term "chaos" as used

113

in mathematics), Robert Shaw, and the meteorologist Edward Lorenz. The following year, independently Pierre Coullet and Charles Tresser with the article "Iterations d'endomorphismes et groupe de renormalisation" and Mitchell Feigenbaum with the article "Quantitative Universality for a Class of Nonlinear Transformations" described logistic maps. They notably discovered the universality in chaos, permitting an application of chaos theory to many different phenomena. [Then] In 1987, Per Bak, Chao Tang and Kurt Wiesenfeld published a paper in *Physical Review Letters* describing for the first time self-organized criticality (SOC), considered to be one of the mechanisms by which complexity arises in nature." (W) (10)

Bak, Tang, and Wiesenfeld's (BTW) paper showed clearly how the existant conditions in dynamic, chaotic systems can spontaneously result in such systems' reaching a point of criticality, that is, a point where a tiny change in conditions can trigger a massive change in the system, an avalanche, in the terms of their experiments. A critical point in a system is, for example, a container of water at 0° centigrade. It can be in the form of a liquid or a solid. It is critical because a small change in conditions can cause it to go either way, in this case, a change in the energy level of the system.

It is not always easy to tell from data whether a physical or other observed process is simply random or chaotic, because in practice no time series consists of a pure "signal". There will always be some form of corrupting noise, even if it is present as round-off or truncation error. Thus any real time series will contain some randomness. In the BTW model, the assumption is made that every such system contains noise which they refer to as 1/f, or 'flicker noise.' and that such systems exhibit a structure with scale-invariant, self-similar (fractal) properties. Some examples of such systems include species-interdependent ecologies in nature, organized ...(from the BTW paper)

"[. . .] such that the different species support each other in a way which cannot be understood by studying the individual constituents in isolation. The same interdependence of species makes the ecosystem very susceptible to small changes or noise. However the system cannot be too sensitive since then it could not evolved into its present state in the first place going to this balance. We may say that such a system is "critical." This qualitative concept of criticality can be put on a firm quantitative basis."
"Such critical systems are abundant in nature. The dynamic of a strict critical stage has a specific temporal fingerprint, namely "flicker noise." ...
"Flicker noise has been observed, for example, in the light from quasars, the intensity of sunspots, the current through resistors, the sand flowing through an hourglass, the flow in rivers such as the Nile, and even stock exchange price indices. All of these may be considered to be extended dynamical systems." . . .
"Despite the ubiquity of flicker noise its origin is not well understood. Indeed one may say that because of its ubiquity no proposed mechanism to data can lay claim as a single generalized general underlying root of 1/f noise. (In this sense) Flicker noise, is in fact not noise but reflects the intrinsic dynamics of self organized critical systems. Another signature of criticality is spatial self similarity. It has been pointed out that nature is full of self similar "fractal" structures, though the physical reason for this too is not understood. Most notably the whole universe is (in one theory) an extended dynamical system where a self similar cosmic string structure has been claimed. Turbulence is a phenomenon where self-similarity is believed to occur in both space and time." (11)

Where does 1/f noise come from? The 1/f or "flicker noise" referred to by BTW is what electronics and sound engineers call "pink noise," one of several "colors" (white, purple, brown, (for brownian), and others) resulting from different factors impinging on, in particular, electronic communication, audio, and video systems. "Pink" noise is a signal or process with a power spectral density inversely proportional to its frequency, or 1/f. It was first discovered in vacuum tubes in 1925 and since then it has been found everywhere from fluctuations of the intensity in music recordings to human heart rates and electrical currents in materials and devices. As Bak et al say, the "pink" variety is ubiquitous, appearing everywhere, requiring special filters to damp its effects on cell phone transmission, for instance.

"After almost a century of investigations, the origin of 1/f noise in most of material systems remained a mystery. A question of particular importance for electronics was whether 1/f noise was generated on the surface of electrical conductors or inside their volumes. " 12)

If we are right in our concept of the ether, however, we know that this "flicker noise" exists throughout the system. It is also significant for our argument, particularly, because one of the key "discoveries" in support of the big bang, the CMBR or cosmic microwave background radiation, those published images drawn from research into the distant reaches of the universe, have been created and "smoothed" by carefully filtering out the 1/f flicker noise, or "pink" noise.

"One thought is that 1/f noise could be related to the spatial structure of matter. Systems in space have many degrees of freedom; one or more degrees of freedom is associated with each point in space. The systems had to be "open," that is, energy had to be supplied from outside, since closed systems in which energy could not be supplied would approach an ordered disordered equilibrium state without complex behavior. However, at that time there were no known general principals for open systems with many degrees of freedom." (13) (Bak, *How Nature Works*)

But there were. *Degrees of freedom* refers to the number of options an object has for movement in space. Once more, from Wikipedia:

In physics, a **degree of freedom** is an independent physical parameter, in the formal description of the state of a physical system. The set of all dimensions of a system is known as a phase space, and degrees of freedom are sometimes referred to as its dimensions. **The position of a rigid body in space is defined by three components of translation and three components of rotation, which means that it has six degrees of freedom.** (W) (14) (emphasis added)

Bertalanffy's *General Systems Theory* and other systems dynamics studies had already shown a necessary "open system" structure for proper functioning in nature *and* theory. It had always been known that the Second Law of Thermodynamics was only true for closed systems, and even there self-organization had been seen to occur. Common sense would also tell us that critical systems must be inherently unstable., but BTW's and other studies have shown us that they are a natural, spontaneous result arising out of chaotic dynamic systems. As Allen B. Downey says in his book *Think Complexity* :

"Critical systems are *usually* unstable. For example, to keep water in a partially frozen state requires active control of the temperature. If the system is near the critical temperature, a small deviation tends to move the system into one phase or the other.
Many natural systems exhibit characteristic behaviors of criticality, but if critical points are unstable, they should not be common in nature. This is the puzzle Bak, Tang and Wiesenfeld address. Their solution is called self-organized criticality (SOC), where "self-organized" means that from any initial condition, the system tends to move toward a critical state, and stay there, without external control." (15)

In the BTW study, the metaphorical example used to illustrate these principals was that of a simple pile of sand. The pile is created by dropping one grain of sand at a time onto a surface, each grain representing a unit of energy. As the pile grows a point, or several points, will occur when the addition of one more grain will trigger an event dubbed an avalanche. The addition of a single grain will disrupt the apparently stable sandpile out of its critical state which it had reached spontaneously and force it into a similar but different stable state, no longer critical, that is, no longer subject to catastrophic change by a small difference in its energy level. "Catastrophic" here does not imply disaster or failure. Rather, it refers to a significant large-scale change in the overall configuration of the system, usually to a new state of "temporary" stability. Another important point to remember is that the occurrence of the avalanche could not and can not be predicted either as to time or location, only that it will occur.

One may see where we are going with this. Ignoring the "string structure" reference quoted earlier as the non-testable, non-provable speculation called "string theory," one can see the universe, and in our case, the electromagnetic field that surrounds and permeates it, as an extended, dynamical, turbulent system, characterized also by the presence of 1/f or flicker noise, within which, in the words of BTW, spontaneous, self-organizing criticality occurs. Order arises, not directly out of chaos, but out of a dynamic system at a point of criticality, a state at which the system has arrived at spontaneously. The first stage of this order is likely to have occurred as a result of another physical process we know and understand well in many other settings. That is the seemingly spontaneous appearance in noisy, chaotic environments, of order, structured patterns, and stable phenomena as a result of reverberation, reinforcement, and resonance

Markus J. Aschwanden, in *SOC Systems in Astrophysics*, (16) opens his paper with, "The universe is full of nonlinear energy dissipation processes, which occur intermittently, triggered by local instabilities, and can be understood in terms of the self-organized criticality (SOC) concept." Aschwanden goes on to show these processes at work in galaxy formation Blazars, soft gamma repeaters (emissions from neutron stars), pulsars, cosmic ray emissions, even in the distribution of lunar craters, the asteroid belt, and the rings of Saturn. The universality of SOC processes is becoming well-documented. Aschwanden Opens his chapter on theoretical models of SOC with a significant question:

"How can the universe start with a few types of elementary particles at the big bang, and end up with life, history, economics, and literature? The question is screaming out to be answered but it is seldom even asked. Why did the big bang not form a simple gas of particles, or condense into one big crystal?. The answer to this fundamental question lies in the tendency of the universal evolution towards complexity, which is a property of many nonlinear energy dissipation processes. Dissipative non- linear systems generally have a source of free energy, which can be partially dissipated whenever an instability occurs. This triggers an avalanche-like energy dissipation event above some threshold level. Such nonlinear processes are observed in astrophysics, magnetospheric physics, geophysics, material sciences, physical laboratories, human activities (stock market, city sizes, internet, brain activity), and in natural hazards and catastrophes (earthquakes, snow avalanches, forest fires)." A tentative list of SOC phenomena with the relevant sources of free energy, the physical driver mechanisms, and Instabilities that trigger a SOC event are listed below." (16)

Table 2.1: Examples of physical processes with SOC behavior.

SOC Phenomenon	Source of free energy or physical mechanism	Instability or trigger of SOC event
Galaxy formation	gravity, rotation	density fluctuations
Star formation	gravity, rotation	gravitational collapse
Blazars	gravity, magnetic field	relativistic jets
Soft gamma ray repeaters	magnetic field	star crust fractures
Pulsar glitches	rotation	Magnus force
Blackhole objects	gravity, rotation	accretion disk instability
Cosmic rays	magnetic field, shocks	particle acceleration
Solar/stellar dynamo	magnetofriction in tachocline	magnetic buoyancy
Solar/stellar flares	magnetic stressing	magnetic reconnection
Nuclear burning	atomic energy	chain reaction
Saturn rings	kinetic energy	collisions
Asteroid belt	kinetic energy	collisions
Lunar craters	lunar gravity	meteroid impact
Magnetospheric substorms	electric currents, solar wind	magnetic reconnection
Earthquakes	continental drift	tectonic slipping
Snow avalanches	gravity	temperature increase
Sandpile avalanches	gravity	super-critical slope
Forest fire	heat capacity of wood	lightening, campfire
Lightening	electrostatic potential	discharge
Traffic collisions	kinetic energy of cars	driver distraction, ice
Stockmarket crash	economic capital, profit	political event, speculation
Lottery win	optimistic buyers	random drawing system

3. Resonance

THE PHYSICAL NATURE OF THE FREQUENCY VIBRATION of an electromagnetic field, in its representations in the literature, follows certain conventions. For example, it is often shown as connected sinusoidal waves at right angles to each other as was shown before.

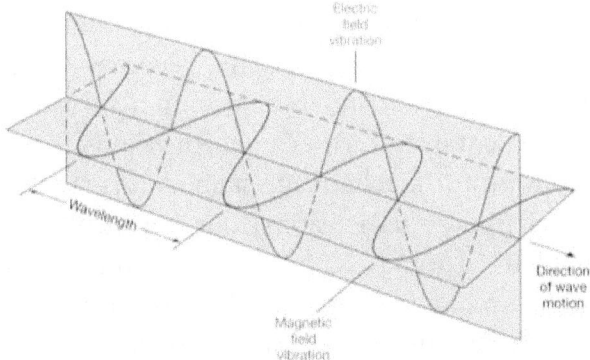

In this diagram the "center" line, the axis about which the vibrations occur, might be considered the null position for positive versus negative polarity, the "zero" position in terms of energy density or magnitude, or something similar. The vibrations of electromagnetic energy in a field may actually be of the nature of pulsations from a negative to positive value radiating from a point. What we do not know is the actual physical nature of that vibration, only that we can capture its pulses on our instrumentation or on our oscilloscope screens which translate its variations into a pre-determined sinusoidal graphical form. We will undoubtedly never actually "see" it. If we could, it might look more like this more fanciful representation.

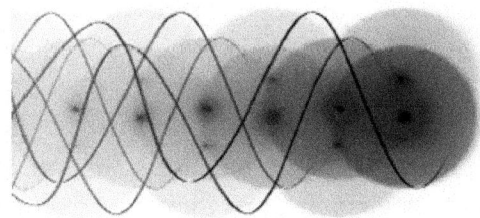

But even though we can't *see* it, like most all of other electromagnetic manifestations, we can detect it in other ways and make practical use of it. So if we seek to determine how elec-

tromagnetic energy in the field of the cosmos concentrates itself, reinforces itself in concentrated points, becomes a temporarily stable entity able to exert its influence on a surrounding
region and is able to catalyze further concentrations in its vicinity, we need to look for mechanisms such as this that work
here in the zone of middle dimensions, our perceptible environment where what Thomas Kuhn calls "normal science"
takes place. There are at least three of those areas, those
mechanisms that will help us in these demonstrations.

The first of these is something we all know and some understand. It is the simple effect we call resonance, a variety of
sympathetic reverberation, or reinforcement. Typically called
interference, it can both positive and negative effects, constructive or destructive. A diagram of how this works in a
wave-like environment is shown here and before.

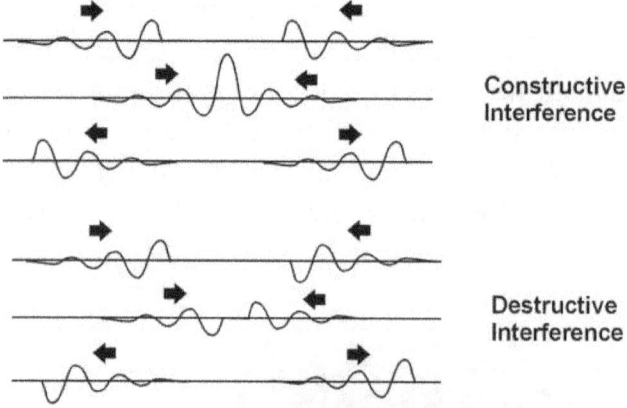

Constructive
Interference

Destructive
Interference

When two waves of similar form and frequency encounter
one another, depending on the particular phase and location
they find themselves in at that instant, they may reinforce one
another or may cancel one another. If three or more happen
to be there, in the same phase, a larger reinforcement will occur. the resultant wave will have a higher energy level, a
greater amplitude. In an environment as vast and extensive as

we suppose the electromagnetic ether to be one can imagine this phenomenon occurring at billions of times and locations. If a sufficient number of sympathetic reinforcements occur at a given point it can and probably will generate an unusually high energy focus sufficient to result in a disruption of the critical condition in that region of the cosmos, resulting in what the sandpile model of SOC calls an avalanche, a region of temporarily stable form, a concentrated energy focus, not unlike one of James Clerk Maxwell's electromagnetic vortices, but without the need for a mechanical interpretation.

In simple English, we would say that the reverberation and its reinforcement had generated a region of resonance, what we know here in the ZMD in the world of sound waves as a temporarily stable pitch. In a random mixture of vibration, certain combinations will have a greater tendency to reinforce one another than others. We typically find those in sound waves to be those with a numerical ratio of additives and multiples in frequency. We call these potentially resonant frequencies or harmonic relationships.

"Resonance is the tendency of a system to oscillate with greater amplitude at some frequencies than at others. Frequencies at which the response amplitude is a relative maximum are known as the system's **resonant frequencies**, or **resonance frequencies**. At these frequencies, even small periodic driving forces can produce large amplitude oscillations, because the system stores vibrational energy.

Resonance occurs when a system is able to store and easily transfer energy between two or more different storage modes (such as kinetic energy and potential energy in the case of a pendulum). However, there are some losses from cycle to cycle, called damping. When damping is small, the resonant frequency is approximately equal to the natural frequency of the system, which is a frequency of unforced vibrations. Some systems have multiple, distinct, resonant frequencies."(W)*

Resonant systems can be used to generate vibrations of a specific frequency (e.g., musical instruments), or pick out specific frequencies from a complex vibration containing many frequencies. occurs widely in nature, and is exploited in many manmade devices. It is the mechanism by which virtually all sinusoidal waves and vibrations are generated. Many sounds we hear, such as when hard objects of metal, glass, or wood are struck, are caused by brief resonant vibrations in the object. **Light and other short wavelength electromagnetic radiation is produced by resonance.** (emphasis added)(W) (17)

Resonance as a phenomenon and as an actor in the physical world shares an important quality with "pink noise" and with self-organizing criticality. It is found everywhere in our world and beyond it. Remember the observations of Balthasar van der Pol, where different coherent resonant frequencies resulted from the increase in the energy of a simple vacuum tube system. We now can see the possible origin of coherent stable concentrations of energy in our electromagnetic field, the new, new ether. Is it possible that these can be the beginnings of cosmic evolution?

2.2 *Cosmic Evolution*

IT MAY SEEM TO BE STRETCHING A POINT to dub these energy foci as proto-universes, but that is, in fact, what they are, stable concentrations of energy in the field, with the potential to increase in size and complexity, entities that could be identified by an observer, if there were an observer. But we know it is all energy. So we have a point of high energy in an electromagnetic field, the result of which in classical physics terms would generate a new field. But since this point, this energy focus, is already in and part of the field, what it actually generates is a region of distortion in that field, a higher energy region whose density falls off as the square of the distance from its focus. And within this newly stable region of higher energy, the probability is increased that additional reinforcements, resonances, even avalanches will occur.

What do we know, in our everyday world, that might be like this? Something that we believe began as a tiny, simple, perhaps even comprehensible entity, but resulted eventually, at least up until now, into something amazingly complex and interesting. Of course we mean that other enduring mystery, that of the origin and evolution of life here on earth. This is, of course, a mystery that we can come close to, can examine in detail, develop theories about, and test those theories in real life, in the field, and in the laboratory. In this quest, we have samples of all sizes and complexity, from very simple on up to ourselves to examine and test. And as a result a reliable body of theory has itself evolved.

Our most complete and reliable theory is that of Charles Darwin, who looked at the world in detail and endeavored to describe how it came about in his *Origin of Species* and following works. Darwin was able to show how the incredible variety of life could have come about, and the likely method of that process. Like some theories in physics, Darwin's theory of evolution has become the de facto Standard Model in the biological sciences. But Darwin did not speculate on how it all got started. What he and his successors did do was to show how, once life had begun, it was able to continue and become more complex and survive and modify itself to respond to changing conditions in its environment. Beginnings were left to theoretical speculations, and will probably remain so.

We do have some ideas of the nature of the environment out of which life arose, some four billion years or so ago. Our chemists and archeologists can show us the likely chemical makeup of the world at that time from evidence both buried and on the surface. It was a sea and an atmosphere rich in a few elements, mainly carbon, oxygen, and hydrogen, probably in the form of carbon dioxide and methane and water, all elements that make up most of our bodies today. And of course that atmosphere and those seas were turbulent and constantly receiving energy from the sun, both directly and from that turbulence. So how did this mix become alive?

Paleontologist Andrew Knoll, a professor of biology at Harvard and author of *Life on a Young Planet: The First Three Billion Years of Life.*, tells an interesting story:

"There was a famous experiment done by Stanley Miller when he was a graduate student at the University of Chicago in the early 1950s. Miller essentially put methane, or natural gas, ammonia, hydrogen gas, and water vapor into a beaker. That wasn't a random mixture; at the time he did the experiment, that was at least one view of what the primordial atmosphere would have looked like.
Then he did a brilliant thing. He simply put an electric charge through that mixture to simulate lightning going through an early atmosphere. After sitting around for a couple of days, all of a sudden there was this brown goo all over the reaction vessel. When he analyzed what was in the vessel, rather than only having methane and ammonia, he actually had amino acids, which are the building blocks of proteins. In fact, he had them in just about the same proportions you would find if you looked at organic matter in a meteorite. So the chemistry that Miller was discovering in this wonderful experiment was not some improbable chemistry, but a chemistry that is widely distributed throughout our solar system." *(Interview on NOVA, July 1, 2004) (1)*

So here are two ways in which the basic building blocks of life could have arisen here on earth: out of the interaction of a high energy source and a rich soup of raw materials right here; or by a collision with one or more meteorites out of the thousands that must have struck this planet in its early life. Is one of these sources the preferred one? Probably not, since as we know, it might only take one successful event like this, one

127

successful beginning, to initiate a process we know must have worked, else we wouldn't be having this discussion.

So life as we know it, most likely arose as an organized entity out of a medium that held all of the necessary elements but were unorganized until a random surge of energy pushed them into an organized, temporarily stable state.

These are conjectures that only recently have found a new champion. Jeremy England, a professor of physics at MIT, has begun to lay out a testable theory that shows how life not only could have appeared in the nonliving universe, but that *"This could mean that under certain conditions, matter inexorably acquires the key physical attribute associated with life."* In the Darwinian sense, the distinguishing characteristic of a living entity is the ability to sustain itself in its environment and to replicate, that is, to reproduce itself. In the article in Quanta that capability is described differently, as:

"From the standpoint of physics, there is one essential difference between living things and inanimate clumps of carbon atoms: The former tend to be much better at capturing energy from their environment and dissipating that energy as heat. . . . England, . . . has derived a mathematical formula that he believes explains this capacity. The formula, based on established physics, indicates that when a group of atoms is driven by an external source of energy (like the sun or chemical fuel) and surrounded by a heat bath (like the ocean or atmosphere), it will often gradually restructure itself in order to dissipate increasingly more energy. This could mean that under certain conditions, matter inexorably acquires the key physical attribute associated with life."

The driving flow of energy — whether from the sun or some other source — can give rise to what are known as *dissipative structures*, which are self-organized by the process of dissipating the energy that flows through them. Russian-born Belgian physical chemist Ilya Prigogine won the 1977 Nobel Prize in Chemistry for his work developing the concept. All living things are dissipative structures, as are many non-living things as well — cyclones, hurricanes and tornados, for example. Without explicitly using the term "dissipative structures," the passage above went on to invoke them thus:

Snowflakes, sand dunes, tornadoes, stalactites, graded river beds, and lightning are just a few examples of order coming from disorder in nature; none require an intelligent program to achieve that order. In any nontrivial system with lots of energy flowing through it, you are almost certain to find order arising somewhere in the system. If order from disorder is supposed to violate the 2nd law of thermodynamics, why is it ubiquitous in nature?"

https://www.quantamagazine.org/20140122-a-new-physics-theory-of-life/

There are important gaps in this story, of course. Even something as simple as amino acids in themselves are not enough to generate a self-organizing or self-reproductive system. That came later, but the process had a start, and four billion years of time ahead of it to work out the kinks, and that story can be read in many books and studies. We'll be talking more about self-organizing systems soon. But first let's look at what happens next in Darwin's evolutionary model. What is essential in understanding Darwinian theory is that the events and outcomes of evolution are not purposeful but random. Changes in biological structure occur not in response to environmental pressure but randomly, and may be either positive and reinforcing in nature as to improving survivability, or negative, resulting in a loss of positive benefits. The outcome is that the positive changes result in an increase in the favored species and a decrease in the unfavored ones. The important point here is the randomness of the changes, the mutations. We only know which were which after the event.

Do these processes have a parallel in our view of the cosmos and the emergence of an ordered universe? Our newborn "energy foci," like those few amino acids we spoke about, may, of course, have a very brief existence. It may quickly, or slowly, resolve itself back into the field as have many before it There may have been many proto-universes generated just like these or even some that are more complex, but destined to never reach maturity. Or some that *have* matured. But we know that at least one has made it or we would not be having this discussion.

How might that have happened?

By the end of the 19th century we knew a lot about waves and wave-like phenomena in nature. We knew about wave forms in water, in sound, how they behaved when in reinforcement, in destructive interference. The wave theory of light, and by extension of other forms of electromagnetic radiation, was in its ascendancy, in spite of Newton's belief in its corpuscular nature. We knew about resonance, sympathetic vibration, even the destructive kind. Even well into the 20th century, after Planck's determination that energy needed to be thought in quanta, the wave-like nature of energy phenomena remained, though in somewhat mongrelized form, as one of two alternative characteristics of electrons and other submicroscopic particles. So controversy still held sway in some quarters.

For big bang theorists, the time from its beginnings as a highly energetic, rapidly inflating plasma to the first particles, protons, took all of 1 microsecond, then their fusion into more complex forms, 1/100th of a second, with the first atomic nuclei, only 3 minutes into the process. The process then slowed, with the first actual atoms, those of free hydrogen, requiring 300,000 years, which, incidentally is the limit to how far we can look back, being the place where we can see the so-called "cosmic microwave background radiation." Here is how they see this history:

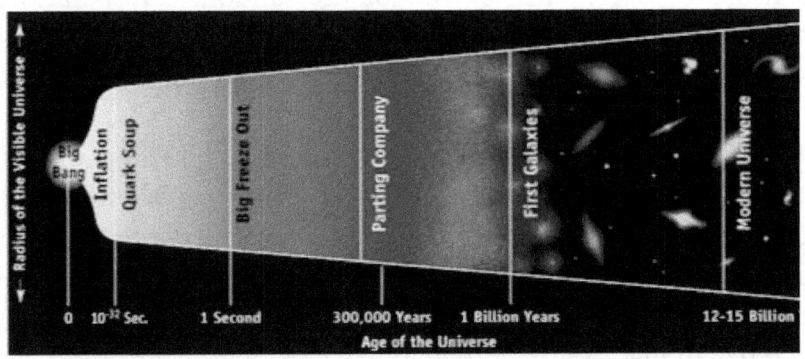

So, if we look back as far as we can see, and add 300,000 years, we have the age of he universe, some 13.7 billion years since it all appeared out of nothing. The only seriously considered alternative theory, that of a "steady state" universe, assumed that the whole of it had instead existed forever, its own difficult to believe justification, and that it appears to be expanding because new stars are being created as we speak, out of something called a C-field, structure unknown, but like many other assumptions of modern physics, assumed to exist because it must, to explain the theory. However, just because something is difficult to believe doesn't mean it couldn't have happened that way. After all the same thing could be said about "Let there be light!" In Biology this would be like saying that when that bolt of lightning hit the earth, or that meteorite, then all of the raw materials for life were created instantly, it all came out of nothing.

If we need a theory that does not require a giant leap of faith, what are we left with? Are there ways in which all that makes up this universe could have come to exist by processes that we either know already or can demonstrate as being possible, even probable? We've done that with the probable

131

origin of the smallest entities, those stable, coherent concentrations of energy arising out of the random reverberations, reinforcements, resonances of waves of energy in our universal energy field, surrounded by their regions of distortion of the field. What is the next step?

Three known processes come to mind. While hints of them and their applicability have arisen for centuries, like the heliocentric theory of Anaxagoras that preceded Copernicus by 2000 years, they were imagined, but in isolation, and their broad applicability only came to full flower in this century. We're talking about the phenomena of phase transitions, the discovery of the fractal geometry of nature, and the concept of cellular automata. Each has been demonstrated to function in broad regions of nature. Each has been shown to be scale-independent, that is, they appear to work similarly and be demonstrable, and to occur at the smallest scale of phenomena up to the largest. And because of this, they share at least the possibility of universality.

"Universality," as Per Bak points out, is the theorist's dream come true. "If the physics of a large class of problems is the same, this gives him the option of selecting the simplest possible system belonging to that class for detailed study." When a researcher, even one who sees himself as a generalist problem solver, sees recurring patterns in different, sometimes even apparently unrelated disciplines, his eyes light up and he imagines the imminent approach of an "Aha!" moment. To the susceptible, of course, universality can be a cruel mistress, leading one away from the prize, into seeing patterns for patterns' sake and down fruitless paths. In Bak's own field, self-organized criticality (SOC), for instance, work published as recently as July 2012 documents strong evidence of its presence throughout the universe.

In the fields we are discussing here the connections and similarities arose late, discovered by other researchers as links between mature disciplines, already well-accepted. A common thread is their tendency toward universality.

1. The geometry of reality—fractals and recursive structures

"Space, in the simple terms preferred by reality physicists, is the seemingly boundless volume surrounding us in which real objects manifest themselves. It is intrinsically empty, inert, 3-dimensional (height, breadth, and depth), and featureless. We ascribe no virtue to space other than its potential to contain phenomena." (Ratcliffe) * (2)

As such, it has no *geometry*, which is in fact simply the name of a system of descriptors and measurements by which we apprehend and communicate information about the objects, events and phenomena that make up the real world we inhabit.

There have been invented numerous geometries, the original to most of us and most complete, is probably that of the Greek mathematician Euclid in about 300 BCE, a study most of us learned in about our first year of high school mathematics. To normal human perception, as in ancient Egypt, the earth appeared flat or mostly so, so measurements and descriptions of the land resembled and were described as on a flat plane, and Euclid's geometry assumed the same. These assumptions led to the axioms that the three angles of a triangular shape added up 180°, two parallel lines would never meet, and so on. But the Greeks knew that the earth was spherical and its surface was therefore curved, so a long set of "parallel" lines eventually met, and a triangle drawn on the earth's curved surface had a total angular value greater than 180°.

As we have pointed out before Mathematics is a wonderful, smooth abstraction of reality, and geometry is no exception. When we began to measure distances and object, of greater size and magnitude than those close at hand, we had to take the earth's curvature into account so new systems of geography had to be devised. And, skipping over a few hundred years of time, when we began to try to measure the elements of the cosmos, we naturally adopted those systems that were in use on earth. "The orb of the Earth, it was reasoned, is mirrored by the celestial sphere, and a horizontal line, strictly speaking, is curved." (Ratcliffe) (3)

As we have also pointed out, the language of mathematics, by the time of Planck and Einstein, had become, ipso facto, the language of physics, with the result that if something seemed to be proven mathematically, then it was assumed to be true in the real world. So true, in fact, that in many cases, observational evidence to the contrary was either explained away or discarded as probably in error. Key examples discussed here include the assumed curvature of spacetime in Einstein's General theory of Relativity; the accelerating expansion of the universe as a result of the big bang and its initial expansion; and the idealized "smoothness" of reality deriving from its mathematical description.

Observation tells us a different story. We know you cannot curve entities that have no real physical attributes, entities like space and time. The assumed expansion of the universe has been challenged, even contradicted, by hundreds of astronomical observations and by many authorities, even including Edwin Hubble himself, the man whose name is most attached to the theory. And the idea that the natural world is smooth is contradicted by observation on a daily basis by anyone who actually observes it. Broken rocks, the outline of a mountain range, the course of a river, none are smooth or featureless in detail. And they stay coarsely defined as you magnify them even up to higher magnitudes. Geometry, however, that purest of abstract disciplines, had, until late in the last century, no way to accurately describe this inherent *roughness* of Nature. Until the invention, in about 1963, of fractal geometry.

In mathematics, a *fractal* is a set that typically displays self-similar patterns at different scales. In practice, they may be exactly the same at every scale, or *nearly* the same at different scales. Self similarity means almost the same as being recursive, or self-referential, except that a mathematical expression may be self-referential without being a fractal. Leibniz, in the seventeenth century, was looking at recursive self-similarity, but limited his thinking to the notion that only straight lines, one-dimensional objects, could be considered self-similar in this way. He referred to them as "fractional exponents" not yet included in any treatise on geometry. A few mathematicians took up these new entities, but little was really accomplished until the mathematician Benoit Mandelbrot happened onto them in the 1960's. Part of the problem was that once one tried to express their character graphically, pencil and paper turned out to be a very tedious and difficult process. The advent of the digital computer changed all that. Wikipedia describes Mandelbrot's arrival on the fractal stage this way:

"In 1963, Benoît Mandelbrot found recurring patterns at every scale in data on cotton prices. Beforehand, he had studied information theory and concluded noise was patterned like a Cantor set: on any scale the proportion of noise-containing periods to error-free periods was a constant – thus errors were inevitable and must be planned for by incorporating redundancy. Mandelbrot described both the "Noah effect" (in which sudden discontinuous changes can occur) and the "Joseph effect" (in which persistence of a value can occur for a while, yet suddenly change afterwards). This challenged the idea that changes in price were normally distributed. In 1967, he published "How long is the coast of Britain? Statistical self-similarity and fractional dimension", showing that a coastline's length varies with the scale of the measuring instrument, resembles itself at all scales, and is infinite in length for an infinitesimally small measuring device. Arguing that a ball of twine appears to be a point when viewed from far away (0-dimensional), a ball when viewed from fairly near (3-dimensional), or a curved strand (1-dimensional), he argued that the dimensions of an object are relative to the observer and may be fractional. An object whose irregularity is constant over different scales ("self-similarity") is a fractal. In 1975 Mandelbrot published *The Fractal Geometry of Nature*, which became a classic of chaos theory. Biological systems such as the branching of the circulatory and bronchial systems proved to fit a fractal model. (4)

Even to the experts, fractals are hard to define. It seems to be another example of complexity that fits the, "I can't describe them, but I know one when I see it," model. Generally however, they are thought to meet some or all of several criteria. :

• Self-similarity, which may be manifested as:
1. Exact self-similarity: *identical at all scales; e.g. Koch snowflake*
2. Quasi self-similarity: *approximates the same pattern at different scales; may contain small copies of the entire fractal in distorted and degenerate forms; e.g., the Mandelbrot set's satellites are approximations of the entire set, but not exact copies.*
3. Statistical self-similarity: *repeats a pattern stochastically so numerical or statistical measures are preserved across scales; e.g., randomly generated fractals; the well-known example of the coastline of Britain, for which one would not expect to find a segment scaled and repeated as neatly as the computer generated designs.*
4. Qualitative self-similarity: *as in a time series.*
5. Multifractal scaling: *characterized by more than one fractal di-*

mension or scaling rule

• Fine or detailed structure at arbitrarily small scales. A consequence of this structure is fractals may have emergent properties (related to the next criterion in this list).

• Irregularity locally and globally that is not easily described in traditional Euclidean geometric language. For images of fractal patterns, this has been expressed by phrases such as "smoothly piling up surfaces" and "swirls upon swirls"

• Simple and "perhaps recursive" definitions (W) (5)

The best way to illustrate these principles, however, is probably with illustrations. Here are a few.

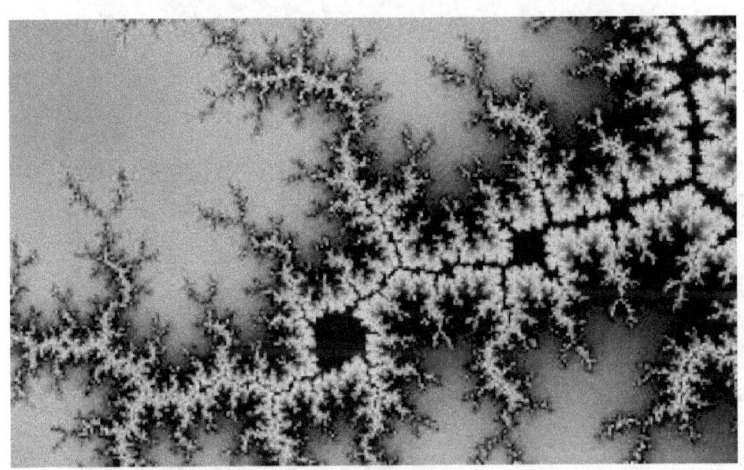

A computer-generated fractal structure.

imagine darkness

A fractal pattern in a coastline

A fractal pattern in a fern.

"Different researchers have postulated that without the aid of modern computer graphics, early investigators were limited to what they could depict in manual drawings, so lacked the means to visualize the beauty and appreciate some of the implications of many of the patterns they had discovered (the Julia set, for instance, could only be visualized through a few iterations as very simple drawings. That changed, however, in the 1960s, when Mandelbrot started writing about self-similarity in papers such as *How Long Is the Coast of Britain? Statistical Self-Similarity and Fractional Dimension*, which built on earlier work by Lewis Fry Richardson. In 1975 Mandelbrot solidified hundreds of years of thought and mathematical development in coining the word "fractal" and illustrated his mathematical definition with striking computer-constructed visualizations. These images captured the popular imagination; many of them were based on recursion, leading to the popular meaning of the term "fractal". Currently, fractal studies are essentially exclusively computer-based." (W) (6)

Back to their universality. As noted above, "Biological systems such as the branching of the circulatory and bronchial systems proved to fit a fractal model." In our sensory system, the accepted number of senses is five, but, as Mandelbrot points out, the actual number of direct sense messages we receive and process is much higher. Remember our discussion of the complexity of the highly complex sound waves we hear in a concert hall. Mandelbrot's original book, *The Fractal Geometry of Nature*, holds hundreds of examples. As Mandelbrot's initial insight in the world of commodities prices illustrates, such patterns permeate economic and social structures, both in humans and primitive species at all levels. Turbulence, as in the atmosphere, exhibits fractal patterns. A pilot flying into a cloud bank discovers that the patterns he saw at a distance are repeated close up as he enters the bank. A difference in scale is indistinguishable. Fractal geometry enables natural systems like human respiration and circulation to maximize their effectiveness while minimizing their footprint, so to speak. It would, on the surface, appear that such systems might have evolved as a result of environmental stresses on populations in biology. but their appearance in non-biological entities such as crystal structures and other complex systems suggests that they might actually be the result of the working out of some rules of organization innate to the underlying structure of the physical organization of the universe. And because these patterns are closely similar at all scales, might not the rules that generate them be closely similar as well? Let's keep this in mind as we look at another recursive model of organization, the class of models that fall under the name of cellular automata. How might these patterns, these self similar structures have arisen?

2. Cellular Automata—The Game of the Universe?

In 1970, in Martin Gardner's column in Scientific American, there was published a version of John H. Conway's "Game of Life." It's effect was to almost instantly popularize the concept of cellular automata as simulations of complex processes in nature. The concept resulted from work done in the 1940s by Stanislaw Ulam and John von Neumann at the Los Alamos National Laboratory. Conway's version was a simple 2-dimensional cellular automaton, its behavior controlled by a simple set of rules. It consisted of a rectangular grid of cells, without borders, in which a cell was "alive" when black, and "dead" when white or empty. A cell's vital status was determined by the conditions in its "neighborhood, the eight cells immediately contiguous. The rules were simple. If a cell's current state was alive, and it had at 2 or 3 live neighbors, it stayed alive; if it had 0-1 or 4-8 live neighbors, it died; if a dead cell had exactly 3 dead neighbors, it came alive; if it had 0-2 or 4-8 dead neighbors, it stayed dead. This behavior is loosely analogous to real cell growth: cells that are isolated or overcrowded die; at moderate densities, they flourish. Depending entirely on the game's starting condition, over multiple cycles, the system may reach a static condition, may die out completely, or may demonstrate almost unlimited complex growth. While manageable on something as simple as a checkerboard, the digital computer allowed it to be operated rapidly. Some examples:

Figure 1, below is an initial pattern known to produce an almost unlimited set of variations.

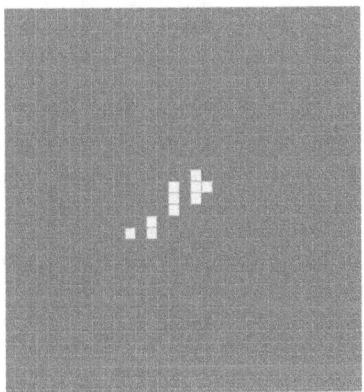

Figure 2 shows the result of the application of Conway's rules after 135 cycles,

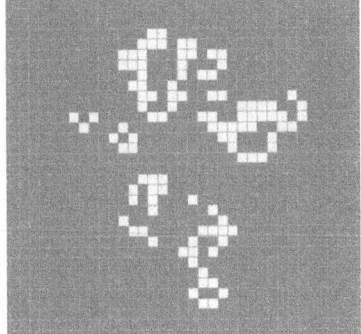

and Figure 3 shows the same pattern after 500 iterations.

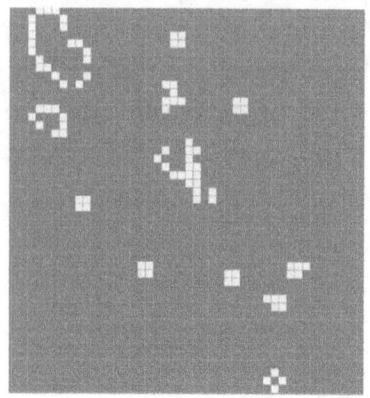

If watched closely one sees certain patterns emerge as seemingly stable patterns, only to be overtaken and transformed by nearby entities, all by reapplication of the same rules and constraints.

Then imagine that instead of the simple two-dimensional grid shown that each square unit represents a single face of a cube, each of its edges "quantized" into 1000 units, resulting in a total of one billion cells for each cubic unit, each one of which is governed by the same rules, and that their rate of vibration is about 1/h, or a frequency on the order of 1.65×10^{35} cycles per second. While the odds of stable outcomes from this process may be very low, the immense number of iterations over time almost guarantees their emergence. If one adds in the factor that as each stable (high energy) point is created, it has the effect of increasing the energy density in its near surround and initiates phase transitions. the possibilities are almost immeasurably magnified.

Stephen Wolfram, known best for his development of the mathematics computer program *Mathematica*, worked extensively with one-dimensional CA's in the 1980s and beyond, ultimately publishing his work as *A New Kind of Science* in 2002, (7) strongly suggesting the use of cellular automata in many scientific fields. Wolfram identified four (behavioral) classifications of cellular automata.

•

• Class 1: Nearly all initial patterns evolve quickly into a stable, homogeneous state. Any randomness in the initial pattern disappears.

• Class 2: Nearly all initial patterns evolve quickly into stable or oscillating structures. Some of the randomness in the initial pattern may filter out, but some remains. Local changes to the initial pattern tend to remain local.

• Class 3: Nearly all initial patterns evolve in a pseudo-random or chaotic manner. Any stable structures that appear are quickly destroyed by the surrounding noise. Local changes to the initial pattern tend to spread indefinitely.

• Class 4: Nearly all initial patterns evolve into structures that interact in complex and interesting ways, with the formation of local structures that are able to survive for long periods of time.

(Ilachinsky)(8)

How does this relate to the real world? After all, this all came out of a desire to simulate reality, not from real world observations. Isn't it just another fairy tale speculation? Well, several scientists and mathematicians are sure that it can be found in physical reality. Marvin Minsky, of artificial intelligence fame, investigated how to understand particle interaction using a 4-dimensional lattice (not one we would think might be fruitful). Edward Fredkin has explored a possible "finite nature hypothesis, the idea that "ultimately every quantity of physics, including space and time, will turn out to be discrete and finite," and hence could be considered CA similar. He and Wolfram share a conviction that CA might ultimately explain physics. Closer to our convictions here is the work of Gabriele Rossi and his collaborators described in

143

The Mathematics of the Models of Reference, that, using a complex CA model might one day shed light on the universe's evolution. Rossi offers a detailed definition of CA-like 3-dimensional processes:

The theory of cellular automata has its roots in the works of prominent logicians, mathematicians, physicists, computer scientists, such as Alan Turing, John von Neumann, Stanislaw Ulam, John Conway, Stephen Wolfram and Konrad Zuse. Cellular automata (CA) are mathematical representations of complex systems. A complex system is a dynamic system - that is, something changing its state over time and reacting to the environment - whose properties are determined by countless feedbacks and interactions between its parts: a cell, the human body, the U.S. economy are all good examples of complex systems whose behavior is often hard to predict. A distinctive sign of CA is their discrete nature, both in space and in time: almost any CA, in fact, has the following five features:
* A discrete lattice of cells (or atoms); the lattice can be 1, 2, 3 or n-dimensional. The cells are the basic, fundamental bricks of the CA
* Homogeneity: each cell is identical to any other cell in the lattice
* Discrete states: each cell can be in one of finitely many possible discrete states
* Local interactions: each cell interact only with a finite number of cells, its neighborhood
    * Deterministic dynamics: at each instant in time, each cell updates its state with a transition function. The state of a cell at a time t only depends on the states of the cell's neighborhood at time t - 1. (9)

The evidence of natural occurrences of CA-like operations in biology and chemical processes is small but it is there.

"Plants regulate their intake and loss of gases via a cellular automaton mechanism. Each stoma on the leaf acts as a cell.

Patterns of some seashells, like the ones in *Conus* and *Cymbiola* genus, are generated by natural cellular automata. The pigment cells reside in a narrow band along the shell's lip. Each cell secretes pigments according to the activating and inhibiting activity of its neighbor pigment cells, obeying a natural version of a mathematical rule.[59] The cell band leaves the colored pattern on the shell as it grows slowly. For example, the widespread species *Conus textile* bears a pattern resembling Wolfram's rule 30 cellular automaton." (10)

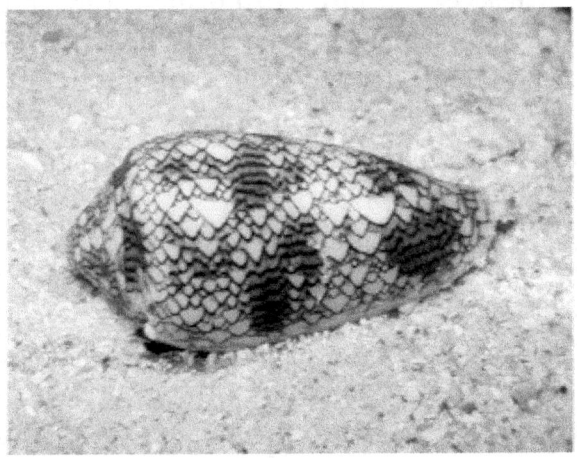

A Textile cone snail (*Conus textile*)
August 2005 Cod Hole,
Great Barrier Reef, Australia
Copyright (c) 2005 Richard Ling

"Moving wave patterns on the skin of cephalopods (octopus, squid) can be simulated with a two-state, two-dimensional cellular automata, each state corresponding to either an expanded or retracted chromatophore, the cells used by some animals as a type of camouflage, called physiological color change or metachrosis. Threshold automata have been invented to simulate neurons, and complex behaviors such as recognition and learning can be simulated.[62]

Fibroblasts bear similarities to cellular automata, as each fibro-blast only interacts with its neighbors." (W)" (10)

Closer at hand and to our hearts is the ability of these models to show how great complexity can be achieved out of apparently chaotic systems by the workings of a few simple rules or instructions. We have seen how this works in biology, DNA being the almost ultimate example of such a mechanism, protein folding another. Swarming and flocking behaviors in large populations of insects and birds also demonstrates how simple rules of local relationships results in large dynamic patterns. If, as we expect, the creation of the fractal nature of reality is itself the result of the application of a few simple rules, we are much closer to determining what those rules might be, and creating "the game of the universe."

The third process that contributes to our theory of processes that lead to the development of larger complex entities and systems out of the tiny energy concentrations arising from resonances in the field, is well known, but only recently better understood, and even more recently become a candidate for the label of universality, and that is the phenomenon of phase transitions.

3. Phase Transitions—a change in state but not in essence

We all learned in high school chemistry about how ice turns to water and water turns to steam, did we not? Here it is in simple terms. Say you have a gram of ice at -10 degrees Centigrade and add heat slowly. Adding ten calories of heat will raise the temperature of that gram of ice to 0° C. but it will still be ice. Now if you continue to add heat, it will remain frozen and at 0° until you have added 80 more calories. Then it will turn to liquid water. That amount, 80 calories, is sometimes called the heat of liquefaction of water (H_2O), and its addition to the system has initiated what is called a phase transition, a change in the physical structure of the ice/water, but not a change in its essence, that of being the same chemical compound, H_2O. Now, continue to add heat to the gram of what is now water, and its internal temperature will rise at the rate of one degree per calorie until it reaches 100° C. It still remains liquid. Continue to add heat and its temperature will stay at 100° until you have added, this time, 540 calories to the system. At that point the water will vaporize into steam. If you have watched a pot of water as it begins to boil, (it does, even if you watch!), you will see this process occurring unevenly. Bubbles will appear and rise to the top even as much of the water in the pot seems not to change. The important point to remember however, is that the substance inside the bubble is the same as that outside of it. Both are H_2O, only inside the bubble another phase transition has occurred.

Phase transitions are ubiquitous in nature. The one we have just described is happening continuously before our very eyes. Water is taken up from lakes, rivers, oceans into the atmosphere as vapor and is returned to us as dew, or rain. In cold climates it regularly is turned to ice sometimes in the form of snow. The steel in your family car's bumper is strengthened in the foundry by undergoing a phase transition in its crystal structure. A few others, from a list in Wikipedia:

The transitions between the solid, liquid, and gaseous phases of a single component, due to the effects of temperature and/or pressure:

A eutectic transformation, in which a two component single phase liquid is cooled and transforms into two solid phases. The same process, but beginning with a solid instead of a liquid is called a eutectoid transformation.

A peritectic transformation, in which a two component single phase solid is heated and transforms into a solid phase and a liquid phase.

A spinodal decomposition, in which a single phase is cooled and separates into two different compositions of that same phase.

Transition to a mesophase between solid and liquid, such as one of the "liquid crystal" phases.

The transition between the ferromagnetic and paramagnetic phases of magnetic materials at the Curie point.

The transition between differently ordered, commensurate or incommensurate, magnetic structures, such as in cerium antimonide.

The martensitic transformation which occurs as one of the many phase transformations in carbon steel and stands as a model for displacive phase transformations.

Changes in the crystallographic structure such as between ferrite and austenite of iron.

The emergence of superconductivity in certain metals and ceramics when cooled below a critical temperature.

The transition between different molecular structures (polymorphs, allotropes or polyamorphs), especially of solids, such as between an amorphous structure and a crystal structure, between two different crystal structures, or between two amorphous structures.

One can add to this the phenomenon of magnetization of some metallic substances resulting from the turbulent behaviors of the fluid cores of planets and atmospheric systems.

In the Standard Models of modern physics and cosmology, the following two unobserved processes have been considered to be in the nature of phase transitions:

Quantum condensation of bosonic fluids (Bose–Einstein condensation). The superfluid transition in liquid helium is an example of this.

The breaking of symmetries in the laws of physics during the early history of the universe as its temperature cooled. (11)(W)

The article goes on to explain the technical process, amazingly similar to the definition of criticality:

At the phase transition point (for instance, boiling point) the two phases of a substance, liquid and vapor, have identical free energies and therefore are equally likely to exist. Below the boiling point, the liquid is the more stable state of the two, whereas above it the gaseous form is preferred. (11)(W)

In some of these processes, the change is totally reversible. The removal of heat from the system causes a regression to the original state. If the steam is kept above the critical temperature, however, it will be stable in that environment. Some of the examples above are, however, permanently changed and remain unchanged unless subjected to another transition process. Metals can be demagnetized for instance by heating above a critical temperature.

It was only in 1971 that the physicist Kenneth Wilson devised a theory of how these processes occur. He showed that the critical phenomena common to phase transitions of many if not all types were the same or closely related and that they could be measured and predicted with confidence. "It (his theory) implies that many systems, different and completely unrelated, can show identical behavior near the critical point. As examples we can mention that liquids, mixtures of liquids, ferromagnets, and binary alloys show the same critical behavior. Experimental and theoretical work from the sixties suggested this form of universality, but Wilson's theory gave a convincing proof from fundamental principles." (12) (W)

Wilson received the Nobel Prize for this work in 1982. Phase transitions, from Wilson's theory as well as common sense, can be seen to occur in recognizable regions of criticality. Systems at criticality can be seen almost universally as having a fractal geometry. The observed progress of forest fires can be seen as having both SOC and fractal characteristics.

1/f noise is found not just in fields but in objects, events and phenomena across the full spectrum of reality, One additional observation: in the Aschwanden paper on *SOC Processes in Astrophysics* cited earlier in our discussion of self-organized criticality makes clear, SOC and criticality in general are closely linked to fractal geometry. In his conclusion, he points out that his thesis is based on:

"[. . .] the fractal-diffusive self-organized criticality (FD-SOC) model in terms of four fundamental parameters: (i) the Euclidean dimension S, (ii) **the fractal dimension DS of the spatial SOC avalanche structure,** (iii) the diffusion index β that includes both sub-diffusion and super-diffusion, and (iv) the energy-volume scaling law with powerlaw index. (Aschwanden) (emphasis added) (13)

One further comment. The energy levels required to initiate phase transitions can be many multiples of those needed to effect other changes in conditions. They can be seen to be many orders of magnitude greater to give rise to a new level of organization. Ice to liquid water requires a "heat of liquefaction" of 80 calories, liquid water to steam (water vapor) requires 540 calories. These relations may be periodic or harmonic. The way in which "forces" are dealt with in the standard models is illustrative. If the so-called "strong force" is designated as 1, the "weak force" is then 10^{-6}.

* * * * *

Herbert Simon offered his well-structured vs. ill-structured dichotomy in a 1973 paper "The Structure of Ill-Structured Problems" (Artificial Intelligence 4: 181-201. In it he gave clarity to a concept hat had plagued problem solvers for generations in nearly every field of endeavor. Not every problem is amenable to narrow linear solutions. Nor is every problem solvable to a single uncluttered answer. Granted, some problems appear to be well-structured. for example, in mathematics one can say "a+b=c," and if one assigns values to a and b the answer is predictable, and in that system always correct. But there exist problems in design, in science, in economics, in social systems and the like that are clearly ill-structured, in the sense that they do not have clear, predictable answers. There may be many possible answers, or there may be only approximate answers, attainable by successive approximations, or what we noted earlier, some sort of perturbation methodology. The problems making up the realm of complexity science are clearly in this category. And perhaps the largest and most complex of them all is that of understanding the nature and the structure of the universe, not to speak of its origins. We have also seen that the solutions to problems of complexity in the many fields it appears in have shown themselves to be most amenable to the methods developed to examine self-organizing systems, that is, SOC, CA, and Fractal Geometry. The model we have been building here is built using those approaches and we think it has shown itself to be a worthy conceptual box to be considered.

We have shown the universe to be a classic ill-structured problem. Actually this is evidenced by the many possible descriptions of its structure and origins that have been developed and presented over thousands of years. More evidence of its complexity is obvious in the fact that none of these theories and models has proven to be complete, consistent, or even based mostly on the evidence around us. The model we have proposed here is different. It is based not on hypothetical speculation but on observations of reality. It depends not on mystical assumptions but on evidence that can be seen and demonstrated in everyday life and the reality that we exist in. A preliminary summary follows, before we begin to detail how this new model works in the real world.

2.3 *Summary: The Simple Universe*

WHAT ALL OF THIS ADDS UP TO.

In the beginning or maybe at a time nearer the middle or perhaps sometime in the more recent past, perhaps as recently as around fifteen billion years ago or so, Perhaps tens of billions of years earlier, a unique event occurred somewhere in the vast sea of cosmic energy from which we arose. That event, in that electromagnetic field, vibrating at frequencies we still cannot measure but higher than that of the gamma rays we can measure, resulted in the creation of a stable, coherent concentration of energy arising from two or a few or perhaps many of those vibrations synchronizing, resonating, and amplifying to a level high enough to initiate and sustain a phase transition to a stable state. That primeval entity, that proto-universe, perhaps only one of many, survived and flourished so as to become the initiating point of the complex self organized system we now recognize as our universe, a temporarily stable finite complex of billions of planets, stars, galaxies and clusters of galaxies.

Thus far this has resulted in all of the complex turbulences and orderly constructs, including ourselves, that we see, hear, touch, taste, smell, dream and imagine. This has all occurred through random reverberations, reinforcements and resonances in the wave-like nature of the field; then through mechanisms like self-organizing criticality; the rule-based iterative, vibratory behavior of the field, driving the development of complex coherent structures through the occurrence of cellular automata- and network automata-like processes; fractal diffusion, also an iterative, scale-independent, self-similar process; a full range of phase transitions; and gravitational aggregation assisted by energy concentrations in the field, resulting in a universe that can clearly be seen as a self-organized system.

Quite simply, we are postulating that the reality we perceive is a product of and, in fact, results directly from, this sequence of resonances, self-organized criticality, and phase transitions in the field and that the result of these are fractal in structure and appearance. The single fundamental substance from which all perceived reality arises is the energy making up that field. That electromagnetic field is discernable all around us in field distortions and coherent organized energy concentrations that have arisen directly out of it by means of the processes described here.

Our history is surely still not complete, but we can now see our universe as an entity in which these processes continue to work, as witness the appearance in it of new stars and galaxies and its apparent continued expansion into the cosmos out of which it arose and by which it is permeated.

We have teased out the universe's history from our perceptions and imaginings, from the sound of the cosmic surf as the energy of the cosmos fuels its growth (like the sound from the beaches of an atoll in the middle of a mighty ocean, a sound we once thought came to us as the echo of the myth of an explosion that made the universe out of nothing), from our understanding of the workings of the stuff that we perceive and the patterns they follow, and from our knowledge and understanding of the fleeing stars beyond us.

imagine darkness

Part 3. The User Manual—FAQ

"The experiments I am about to relate ... may be repeated with great ease, whenever the sun shines, and without any other appa-ratus than is at hand to every one."
—Thomas Young

imagine darkness

3.1 Introduction

HOW, THEN, DOES THIS THEORY, *the simple universe,* explain the phenomena we have observed in science throughout history? Theorists and scientists, over the last several hundred years, have drawn from their experiments and observations certain conclusions about how nature works. Early on this focus was on the mechanical structure of the universe in both its near and far manifestations. Astronomers since the time of Copernicus through Hubble and his contemporaries showed us how the stars and planets moved and how we might measure those actions and make predictions from them, but were able to offer no explanations of their origins. Newton's explication in the world of matter and forces still informs us here in the ZMD. His mathematical inventions and his studies of light opened up many intellectual doors for the researchers and mathematicians that followed him. He gave us the con-

cept of gravity to replace Aristotle's model of 2000 years before. I am convinced that while the observations of the ancients, even the moderns, may be absolutely correct, may be as
complete and comprehensive as the techniques and instrumentation of whatever their day, but that through misinterpretation, misattribution, and a habit of narrow vision. the conclusions drawn from those observations must be re-examined,
looked at from a different perspective. Because the view from
the old observation post has left us wanting and faced with
too many paradoxes and contradictions.

As Mark Engel puts it in his preface to Gregory Bateson's
book, *Steps to an Ecology of Mind*, "...we create the world
that we perceive, not because there is no reality outside our
heads, but because we select and edit the reality we see, to
conform to our beliefs about what sort of world we live in."
(1)

It is important that we step outside of that "belief" world
and see reality once again, with all its roughness and its warts.
Left to its own devices, the universe is rough—its shapes, its
surfaces, its sounds, its smells, its tastes. Man has sought from
as far back as we can see to idealize the smooth, the sleek, the
precise, the ideal. The language of mathematics has served
those goals well. But we cannot let it be a substitute for reality,
only a descriptor, as Benoit Mandelbrot has shown us.

Part 3 takes up these questions. We will discuss how this
new model, *the simple universe*, describes the world and how
it works, and when necessary will contrast those workings
with the incomplete, unclear, and often contradictory explanations provided by current theories and models. In particular we will show how the desire for mathematical smoothness
has dominated past and current theorizing, and how the real
world must be described in ways that take into account its inherent roughness, its varied and complex recursive nature, its

very different kind of unpredictability from the unique esoteric probability mathematics of Bohr, Heisenberg, and Schrödinger. The inconsistencies, the contradictions, and the essential wrongness of modern physics will come up for discussion.

In this part of our journey we will be exploring everything from the ultramicroscopic world of our most current technology, through the day-to-day regions of the zone of middle dimensions, on out to the cosmos and beyond, including how what we know today is better explained by this new model, in physics, astronomy, and cosmology. We will be touching on the speed of light, the photoelectric effect, general relativity, redshifts, the origins of stars and planets. We will devote some time to refutation of misattributed observations like the so-called cloud chamber tracks of particles, some of the mystical detours of modern physics, the nature and limitations of constants and open parameters. We will discuss old questions like those regarding transparency, translucency, opacity to light, and conductivity, resistance, and insulation as they relate to electricity and electrical currents. We will explain old observations, like atoms, molecules, and the periodic table, by looking at them in this new way. And hopefully, we will point out directions to new laws, of objects, of events, of phenomena, of energy. And in the cosmos, there will be new explanations of the "mysteries" of dark energy, dark matter, expansion of the universe, , the proliferation of heavy elements, and turbulence.

3.2 Microcosmos/—imagining the Unseeable

1. Unsolved mysteries

OCCASIONALLY, EVEN TODAY as I write this, in September 2014, one sees in the popular press a speculative article titled, perhaps, "What are today's most perplexing unsolved mysteries in science?" Today's version asks about the origins of the current scary disease, the Ebola virus, now rampant in a small region of west Africa. Some discuss the unpredictability of weather, today another deals with earthquake prediction in Japan. In modern physics the seemingly longest running candidate, about 100 years-plus so far, is the apparent irreconcilability of the two current standard models, that of particle/quantum physics and its cosmological equivalent the general relativity/ big bang theory, still being debated by many. Others have given up the search, claiming that micro-physics is just bound by different laws and we should just accept that there will never be a reconciliation. The optimists among us share a different belief, however. We find it inconceivable that there should be one set of physical laws at the micro-scale and another at the cosmic limits, or, as some propose, that there are likely to be multiple universes, each with its own set of rules by which it is organized. If you have read this far, you have seen one such approach to a new resolution of this debate, that by simply giving up one simple belief system, the long held conviction of the existence of an invisible entity called a particle, one can see a clear way through to a single set of laws that govern everything from the smallest conceivable scale to the largest.

Absent any other kind of organization, and since the field is so vast, we'll start at the smallest level, that now defined as the "quantum" level. Some of this may seem repetitive, since in the explication of the basics of our theory of the structure, organization and origins of the universe. it has been necessary to discuss the alternatives that have been laid out over the history of our science. We will try not to be too boring in describing them again. At the tiniest scale there is, of course, light. Not only is it the likely finest grained entity we can have direct knowledge of, it is the one on which we depend most fully in our observations of the other phenomena of physics. In discussing light are three important aspects of light that illustrate the important differences between prevailing theories and those of The Simple Universe. These are 1. the constant velocity of light, "c", whose limit no known movable entity can exceed; 2. the photoelectric effect, Einstein's Nobel prize winning explication, and the source of one of our major energy generating efforts; and 3. the differences between transparency, translucency, and opacity, differences we all see, but which few of us actually understand. But first, a general discussion of what light is and its place in the electromagnetic spectrum.

2. "*and there was light!*"

An appropriate place to start the task of building a new model of the universe based on observable reality would seem to be with the natural phenomenon we know as light. It is by means of the light striking our light-sensitive organs of sight, our eyes, that we first became aware of observable reality. It is by means of light that we can discern, distinguish, describe, measure, compare the other phenomena that make up our observable universe as well as all the individual elements that make it up. A huge proportion of the knowledge that makes up our sciences is based on our visual observations. Light, we know, can be projected, transmitted, reflected, refracted, absorbed. We measure its velocity, its frequency of vibration, its energy intensity, its spectral range. We know it to be just a part, a narrow humanly visible portion of the broad spectrum of electromagnetic radiation. We use this knowledge in nearly every field of human endeavor, from life generating, life saving, life enhancing devices and systems. And we use it in life ending and life destroying devices and systems as well. We know a lot about it, and we think we know what it actually is, but we can't seem to agree on that last item. The accepted standard model of physics says that light is both a wave and a particle, and which of those two manifestations we see depends on how and when we look at it in any given situation. And we can't predict which of those it will turn out to be. How did we get into this conundrum?

First of all, let's clarify what we mean when we say "light" in the discussion following. In general, modern physicists accept that visible light, the phenomenon that we use in our observations of the real world, is simply a portion of the full electromagnetic spectrum. So when we discuss the physical characteristics of light, such as its velocity, its frequency, its absorption, its energy, etc. we are usually talking inclusively about all electromagnetic phenomena. Some of those characteristics vary, of course, depending on frequency and wavelength, that is, on where in the full spectrum a particular range of radiation values occurs.

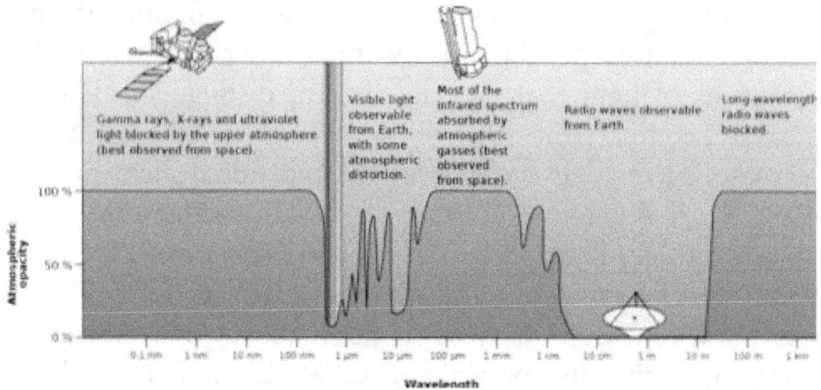

"The **electromagnetic spectrum** is the range of all possible frequencies of electromagnetic radiation.[4] The **visible spectrum** is the portion of the electromagnetic spectrum that is visible to (can be detected by) the human eye. Electromagnetic radiation in this range of wavelengths is called **visible light** or simply light. A typical human eye will respond to wavelengths from about 390 to 700 nm.[1] In terms of frequency, this corresponds to a band in the vicinity of 430–790 THz. (1)

In the 13th century, Roger Bacon theorized that rainbows were produced by a similar process to the passage of light through glass or crystal.[2]
In the 17th century, Isaac Newton discovered that prisms could disassemble and reassemble white light, and described the phenomenon in his book *Opticks*. He was the first to use the word *spectrum* (Latin for "appearance" or "apparition") in this sense in print in 1671 in describing his experiments in optics. Newton observed that, when a narrow beam of sunlight strikes the face of a

glass prism at an angle, some is reflected and some of the beam passes into and through the glass, emerging as different-colored bands. Newton hypothesized light to be made up of "corpuscles" (particles) of different colors, with the different colors of light moving at different speeds in transparent matter, red light moving more quickly than violet in glass. The result is that red light bends (is refracted) less sharply than violet as it passes through the prism, creating a spectrum of colors." (2) (Wikipedia)

The conundrum dates from here. The mind-model of scientists, from earliest times, was based on the obvious perception that the movement of anything, from a running gazelle, to a thrown rock, was a process of penetration though, or, in the case of a ship on the ocean, on and supported by, a medium. In the case of a ship, the water of the ocean. In the case of a falling object or a thrown ball, by the air. We knew the media were there because we could see or feel them and their resistance to the motion of the object. In the case of light, therefore, there must be a medium that carried it from its source to our organs of perception. In Newton's studies, the light was bent as it passed through a medium, the glass of the prism, but was bent more depending on its particular color, so it must be made up of multiple colors. And since this phenomenon of separation occurred as the light passed through a resistant medium, the glass of the prism, the different colors of the light must consist of tiny particles themselves (he called them corpuscles) each with its own intrinsic color and different resistance to the medium. It was not until Helmholtz and Thomas Young's experiments with light in the early 19th century that this concept of the nature of light was seriously challenged.

"In Young's own judgment, of his many achievements the most important was to establish the wave theory of light. To do so, he had to overcome the century-old view, expressed in the venerable Isaac Newton's "Optics", that light is a particle. Nevertheless, in the early 19th century Young put forth a number of theoretical reasons supporting the wave theory of light, and he developed two enduring demonstrations to support this viewpoint. With the ripple tank he demonstrated the idea of interference in the context of water waves. With Young's interference experiment, or double-slit experiment, he demonstrated interference in the context of light as a wave."

"The experiments I am about to relate ... may be repeated with great ease, whenever the sun shines, and without any other apparatus than is at hand to every one." (3)

Through the next century, the nature and behavior of electromagnetic radiation was discovered, studied, and codified, by Herschel, Faraday, and ultimately Maxwell, who firmly established light as a particular part of an infinite spectrum of electromagnetic radiation, all travelling at the same velocity.

Now we know what light is. It's an electromagnetic phenomenon, traveling in or through a medium, at a particular velocity, and behaves and can be detected in the form of waves. A wave, of course, is not a thing in itself. A wave is a "form" of something, as waves in water, or waves in air (sound), for example. So what is light a wave of? We detect it as a wave of energy. But what is its medium?

It's about now, around 1900, that a great deal of time is being spent studying energy, and about this time that Max Planck is coming up with the notion of energy being measurable only in small units he calls quanta. And it's about this time that Einstein and others are looking at the effects of light on other substances. This perfect storm of studies results in Planck's quanta being picked up and applied to light in the invention of the photon. Back to Newton's corpuscles. Back to Lucretious' "uncutable" atoms, and straight ahead into "wave-particle duality." Suddenly light is not either a wave (of something) or a particle. Now light is both a wave and a particle. Or a particle that has a frequency and an amplitude. But definitely, at least a particle, because *all is particles* in the standard model. This is our conundrum.

When Newton saw light as corpuscles, he too saw the need for a medium it must be "passing through," else how could it be resisted and slowed to form the colored spectrum. He then fell back on the idea of Boyle, that of an invisible, undetectable medium he called the "luminiferous ether" named this because it supported light. Others who followed him adopted the same construction, until 1887, when Michelson and Morley's experiment that indicated the non-existence of anything that could be called a particulate medium that might offer resistance to the passage of light. In his Nobel prize lecture, Lorentz proposed an ether that was so fine that it passed through everything and so resisted nothing, but still there was no notion of what it might consist.

With Einstein's theories of relativity, along with a number of experimental verifications, the velocity of light's passage through space was firmly established at 299,792,458 meters per second. It can be slowed, as even Newton knew, but this is a velocity that cannot be exceeded, the ultimate speed limit. And all electromagnetic radiation must follow the same rule. "But wait, suppose I'm riding in a very fast train, racing a beam of light?" (This was the young Einstein's imagining that first set him on the road to his theories). "Wouldn't the speed of light seem slower?" Surprise. It would not seem slower. No matter how fast you are traveling, the speed of light at your side is the same, 299,792,458 meters per second. How can this be?

In *the simple universe*, of course, there *is* an ether, the background (and foreground) electromagnetic field. In *the simple universe* there are no particles, no photons, no quanta, except perhaps as a minimal unit of measurement. Without particles, there is no particulate, resistive medium that light must pass through to get from here to there and back again. In *The Simple Universe*, the phenomenon we call light is in fact an orderly coherent concentration, a patterned distortion in the field of which it is a part, carried, as, here in the ZMD radio signals are carried in the field. It operates just as sound, not as a separate entity passing through a medium, but as a complex distortion of that medium, the atmosphere. And just as the velocity of sound transmission here at sea level is a constant, controlled and limited by the structure of its medium, so is the velocity of light controlled and limited by the structure of the field, and is thus a measurable constant as well.

Can this approach shed light(!) on other experiments as well? Let's look at some of them.

2. *The velocity of light*

The first, and principal example is that of one of our primary physical constants, "c" the velocity of light in a vacuum. The velocity of light is a generally accepted constant, even to its becoming part of the definition of a unit of measure, the meter. Its velocity is established at 299,792,458 meters per second, verified by now probably thousands of experiments. And, importantly, it has been established as a unique standard, one that is totally independent of the theory of relativity. In other words, if an observer is moving rapidly away from a source of light, he does not measure its velocity as slower than c, as one would, for example, in passing a slower moving car on a highway. c is a constant, regardless of the speed of the observer. Light can be slowed, of course, below that maximum limit, when passing through another medium. But for example, when light at the velocity c passes through, say, a thickness of translucent material like glass, its velocity slows while it is in the glass, but resumes its original velocity instantaneously when it leaves. Its total energy has been reduced, so its brightness is less, but its velocity has been instantly restored. The lost energy from its passage is observable as heat, or the electrical phenomena discussed previously, or as a change in the physical structure of the impeding medium. Examples abound.

"The entire universe is made up of waves in a field of energy, but these can sometimes appear to us as if they are particles." This is the premise on which this entire work stands. It follows then, that light, too, is a wave-like electromagnetic phenomenon, not particulate, and that the medium of which it is a part is an electromagnetic field, then its velocity must, of necessity, be determined by the underlying structure of that medium . Note, that I did not say "—through which it passes." There is no resistive medium through which it must force its way, which "massless" photons might find impossible. What we call "light" is an electromagnetic wave phenomenon of a particular range of frequencies, visible to our sense organs and their prosthetic enhancers, which is carried on an underlying wave structure just as our wireless telephone signals and data, our radio communications and the like are carried, at a constant, invariant velocity, unimpeded by molecules of the atmosphere and most other supposedly resistive elements.

What, then, is then, is the likely nature of this medium, and how does it connect to the constant value of the speed of light? The answer seems to lie in that other constant, the one called Planck's Constant. It is a very precise number,

$6.62606957 \times 10^{-34}$ m^2 kg/s. It ties c, the velocity of light, to the total energy of an event or entity. From this relationship Planck developed a set of "Planck values" the most significant of which for us in this discussion is the Planck length, l, a unit of length equal to $1.616199(97) \times 10^{-35}$ meters, which, in most circles of modern physics, is considered to be the shortest measurable length (4). Originally derived from the relationship between Planck's Constant, the velocity of light, c, and the gravitational constant, G, this value turns out to be remarkably close to the value of 1/h, the reciprocal of Planck's Constant itself. The formula which ties observed values to the underlying structure of the universe is $E = h\nu$. In the minimal case the E involved is meant to be the energy of a photon, ν the frequency, and h, the constant derived from observation. This might be seen to imply that a photon is a *quantum*, but that would be a conceptual error. The mathematics serves the same purpose if E refers to it simply as a unit of energy carried by a unit value of a wave.

Planck intuited that physical <u>action</u> could not take on any indiscriminate value. Instead, the action must be some multiple of a very small quantity (later to be named the "quantum of action" and now called Planck's constant). What this tells us is not that everything must be organized as individual "quanta" of energy, but that the underlying structure of the electromagnetic field may be continuous and smooth, but has a lowest limit of measurement, that is, the smallest unit of measure available to us equal to 1 Planck Length.

Trying to measure anything using a smaller scale than this would then be impossible. For example, if one draws a line on a sheet of paper and places on the line tick marks one "Planck length" apart, each segment of the line does not become automatically independent of each of the other units on the line but remains part of its continuous structure, just as a frequency value is not a separate unit of a wave but only a unit of measurement. Taken to the next step, this relation suggests a value for the frequency of vibration of the underlying electromagnetic ether is likely to be 1/h. That frequency would then have a value of about 1.612×10^{35} Hz. The invariant velocity of light that we have observed and measured is a direct clue to this conclusion.

There is a famous comment by the creator of fractal geometry, Benoit Mandelbrot, in his paper "How long is the coast of Britain?". The response to the question is "how short is your ruler?" The smaller the ruler, the more detailed and the longer, the measure will be. What Planck's Constant gives us is the length of the shortest unit of measure we can use to measure the size of the universe.

So, the answer to the question "What limits light speed to 299,792,458 m/s, not more nor less?" is simply that its velocity is inextricably tied to the frequency of the finest level of the "fabric of the cosmos" as some have called it, on which light is carried. and which permeates all of the real fields and objects, those visible and otherwise detectable, coherent, stable distortions of the electromagnetic field that make up the elements of our universe, perceivable to human senses or not, in its vast but definable bounds.

Put quite simply the velocity of light is limited to 299,792,458 m/s because light is not, as we have previously pointed out, a separate substance passing through a medium, encountering resistance as it goes, but is in fact a disturbance in the medium itself and its velocity is governed by that structure just as the velocity of sound is determined, and limited, by its medium, the atmosphere.

3. *The photoelectric effect*

From Andrew Zimmerman Jones:

"In 1905, <u>Albert Einstein</u> published four papers in the *Annalen der Physik* journal, each of which was significant enough to warrant a Nobel Prize in its own right. The first paper (and the only one to actually be recognized with a Nobel) was his explanation of the photoelectric effect." (5)

The photoelectric effect, first documented by Heinrich Hertz in 1887, in *Annalen der Physic,* though not fully explained, is derived from the fact that certain metallic and some non-metallic substances are known to generate electrical current when exposed to light. The keyboard on which this statement is being typed makes use of that effect by means of a photocell in its body, that when exposed to light generates a current sufficient to maintain a rechargeable battery that enables it to transmit information to my computer. Solar cells now in widespread use are similar. Based on the assumption common in 1905 that electrical "current" consisted of the movement of free electrons (particles) through a substance, it was assumed that the energy contained in the light striking the surface was sufficient to free electrons from their chemical bonds in the material exposed to the light. Experimental studies showed that certain colors, that is, certain frequencies of light accomplished this more effectively than others. Classical wave theories of light and other electromagnetic phenomena could not adequately explain this anomaly. Einstein successfully applied Planck's invention of the "quantum" nature of energy, that energy could only be adequately described as tiny packets, not as a continuous substance, to this issue, and postulated the existence of a "light particle" he called a photon. In effect, his answer was that only photons carrying a sufficient level of energy that is, those well above the energy level of "red" light could enable an electron to break free with sufficient energy to produce an observable current. His main contribution to the debate was the assertion that the observed energy transaction could only occur on a "one-to-one" exchange of energy. In other words, one energetic photon would could displace one electron, while even an entire fleet of less energetic photons could not bring this about. Quantum theory is born!

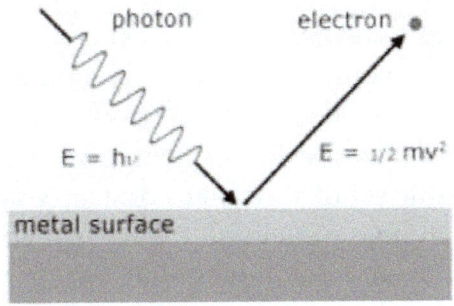

Einstein's paper had two important consequences. The first was very much like that of Michelson and Morley's 1887 study that supposedly dismissed the idea of an ether from any further consideration (it did not). In this case the classical theory of light as a purely wave phenomenon was, in the words of Zimmerman Jones, "crushed." Not completely, of course. No one could deny that light *behaved* in a wavelike manner, but Einstein's paper was assumed to prove that light was made up of particles. Wave-particle duality is born!

An aside, but a fundamental one. This discussion, that of the dual nature of light, illustrates a point that has been referred to many times in this book, too common in modern physics, as it is in politics and most public discourse, that a type of logical confusion appears between differing logical types. In this case the use of the term "particle" is a reference to something presumed to be real, a physical object, so to speak. A "wave" on the other hand, is a description of the state of something, its form, its character. A wave must be a wave of something else, not of another wave. The *duality* here, as in many other aspects of modern physics, is between reality and its description. "Wave" is a descriptor, not an object. So, in this so-called duality, what is the wave a wave of? The logical thing is to say it is a wave of energy, but that is not explicit anywhere in the discussion. Can a logical fallacy endure for over a hundred years? Of course it can!

In *the simple universe*, of course, light is, as we have said, a complex construct of energy. Its behavior is observed as wave-like just as in classical physics. Its interaction with certain "materials" is that of one set of wave-like constructs with another, since the "material object" is itself a wave-like energy construct, a considerably more complex one, of course. The interaction of light with these other complex entities depends on their relative complexity, their energy density and many other factors we must still tease out. Some of the waves of the one will cancel, some will reinforce those of the other entity, the resultant transfer of energy modifying both, but not affecting the fundamental structure of the field. It can be seen that these interactions can explain the transparency of some materials, their translucency, their opacity, all dependent on their internal complexity relative to that of light. In no instance does it require the presence of "electrons" or any other imaginary particulate entities.

4. Transparency, translucency, opacity

These are the observed characteristic behaviors of light that Newton wrestled with almost continuously for many years. How is light affected when it impinges upon or encounters other materials? Why does it appear to be partially or totally blocked in that encounter? Why, in some cases, does it appear to be totally unchanged by its passage through another medium while in others its colors are changed, spread out in an array of different colors? How is it the edges of shadows appear fringed and not as crisp as the edges of the object casting the shadow? How is it that a straight object like an oar appears to bend sharply at the interface between air and water. What is it in the nature of light that can explain all of these phenomena?

One of Newton's predecessors, the Italian physicist Francesco Grimaldi had, in 1660, coined the term diffraction to describe the phenomenon that occurred when light was passed through a narrow slit or through two or three or more slits. It was clear to Grimaldi that the light was being bent around corners, spread out from its narrow slit and illuminating the background with multiple bars of light and dark, just as the water flowing around two rocks in a stream or how sound waves could be heard around corners. Grimaldi's conclusion, that light was wave-like in its manifestations, was published in 1665, two years after his death. (6)

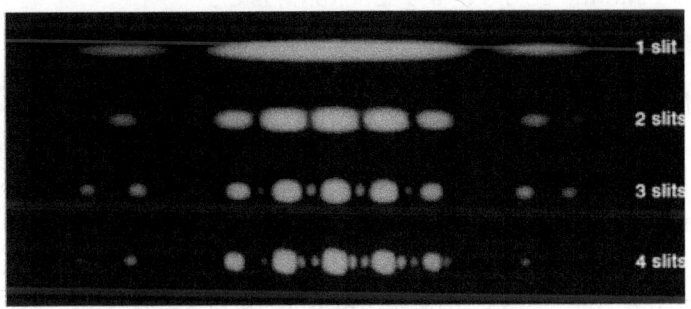

Grimaldi's experiment illustrated

In 1666, Newton's experiments with light showed that ordinary (white) sunlight entering a triangular prism of glass was bent from its entering path but that it then was displayed in a range of colors, suggesting that the entering white light must be made up of all of these colors, but that these were differentially bent while passing through the prism. He further found that if this spectrum as he called were then passed through a lens and then a second, inverted prism, they were reconstituted into a single beam of white light.

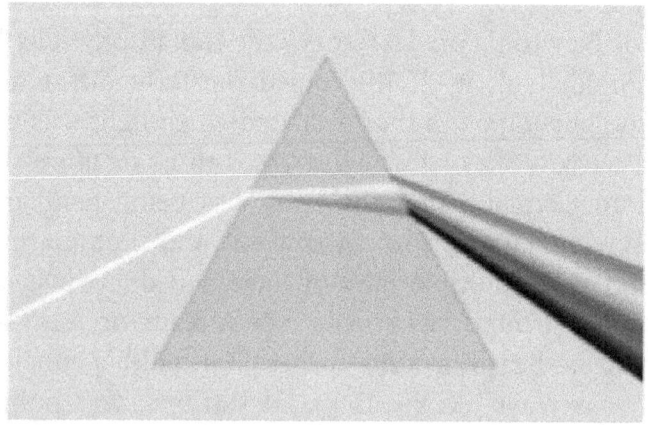

Illustration of a dispersive prism decomposing white light into the colors of the spectrum, as discovered by Newton

In spite of Grimaldi's and others' findings that light was wavelike in its manifestation, Newton remained convinced that light was made up of particles or corpuscles, as he called them and that those of different colors were slowed differentially as they passed through a medium such as the glass of his prisms. He argued that Grimaldi's diffraction was only a different kind of refraction and that their geometric properties could only be explained if light was particulate in nature, since waves did not only travel in straight lines.

(Another aside from Wikipedia:)
"In his *Hypothesis of Light* of 1675, Newton posited the existence of the ether to transmit forces between particles. The contact with the theosophist Henry More, revived his interest in alchemy. He replaced the ether with occult forces based on Hermetic ideas of attraction and repulsion between particles. John Maynard Keynes, who acquired many of Newton's writings on alchemy, stated that "Newton was not the first of the age of reason: He was the last of the magicians." Newton's interest in alchemy cannot be isolated from his contributions to science. This was at a time when there was no clear distinction between alchemy and science. Had he not relied on the occult idea of action at a distance, across a vacuum, he might not have developed his theory of gravity." (7)

Newton's arguments were not unchallenged. Robert Hooke and Christiaan Huygens both published the results of experiments that "proved" that light was wavelike rather than particulate, but as we have said before, those arguments required that the wavelike phenomena be in a medium that they could not yet see or point to but must exist, a "luminiferous ether."

"In 1704, Newton published *Opticks*, in which he expounded his corpuscular theory of light. He considered light to be made up of extremely subtle corpuscles, that ordinary matter was made of grosser corpuscles and speculated that through a kind of alchemical transmutation "Are not gross Bodies and Light convertible into one another, ... and may not Bodies receive much of their Activity from the Particles of Light which enter their Composition?""(Wikipedia, op. cit.)

While these 17th century arguments did not lead directly to those of modern day physics, of photons, quantum mechanics, and the paradox called "wave-particle duality," they foreshadowed them in an interesting way. While Hooke and Huygens, in part, Newton himself, believed in the existence of an ether, modern physics does not. Wave-particle duality in fact turns out to be a verbalizational fallacy. We have pointed out multiple times the problem that threads through almost all of modern physics, that of confusing and conflating *things* (objects, events, phenomena) with the words or formulae that describe them. Korzybski's "orders of abstraction" and Russell's "logical types" show us that a thing and its description cannot be directly equated. A "particle," if it exists, is a thing, an entity. A "wave" is a description of a geometrical characteristic of some material or medium. A wave must be a wave of something. We hear the phrase, the frequency, or the wavelength of a photon, a quality which, if a photon is a particle, it cannot have. If the photon is in vibratory motion in space perhaps, that motion might have a frequency or wavelength as a descriptor, but those are not intrinsic qualities of the entity itself, they are behaviors. I will be accused of verbally quibbling here but it is essential that we carry out this kind of discussion in a common language and agreed-on syntax.

We have learned, of course, that light *can* be made to travel in narrowly focused straight lines. Lasers are ubiquitous in modern technology, and lenses have generated narrow beams for centuries. So Newton's argument that waves could not travel in straight lines was pushing the envelope a bit. We derive straight lines (beams(of light by focusing them, by screening out parts of the incident light. Newton himself reduced the incident light to his prisms by the pinhole or slit approach. The idea of a particle, a photon or corpuscle, is unnecessary to explain refraction. Just as sound radiates in all directions from its initiating source in its medium, so is light in its own medium. Both are, in this sense, compressive and rarefactive phenomena, not transverse waves. The distinction between longitudinal waves and transverse waves becomes moot. Sound *and* light both obey Newton's inverse square law.

In their interactions with light different transparent and translucent substances demonstrate differing behaviors. Light is refracted to a greater degree in say, water, than it is in air. This difference is easily measured and is generally assumed to be the ratio of light's velocity in a vacuum to its velocity in the substance in question. In air, for instance, that ratio is 1.000 293. In water it is 1.333, in soda-lime glass it is 1.46, in a diamond 2.42. The explanation of that behavior at a microscopic scale can be simple or complex. Here, for example, is that description in the terms of the standard model as it appears currently in the Wikipedia article on the index of refraction: (The length of this quote is included here for the purpose of demonstrating the complexity into which we are forced by our adherence to the so-called standard model.)

"At the microscale, an electromagnetic wave's phase velocity is slowed in a material because the electric field creates a disturbance in the charges of each atom (primarily the electrons) proportional to the electric susceptibility of the medium. (Similarly, the magnetic field creates a disturbance proportional to the magnetic susceptibility.) As the electromagnetic fields oscillate in the wave, the charges in the material will be "shaken" back and forth at the same frequency.[14] The charges thus radiate their own electromagnetic wave that is at the same frequency, but usually with a phase delay, as the charges may move out of phase with the force driving them (see sinusoidally driven harmonic oscillator). The light wave traveling in the medium is the macroscopic superposition (sum) of all such contributions in the material: the original wave plus the waves radiated by all the moving charges. This wave is typically a wave with the same frequency but shorter wavelength than the original, leading to a slowing of the wave's phase velocity. Most of the radiation from oscillating material charges will modify the incoming wave, changing its velocity.

However, some net energy will be radiated in other directions or even at other frequencies (see scattering).

Depending on the relative phase of the original driving wave and the waves radiated by the charge motion, there are several possibilities:

If the electrons emit a light wave which is 90° out of phase with the light wave shaking them, it will cause the total light wave to travel more slowly. This is the normal refraction of transparent materials like glass or water, and corresponds to a refractive index which is real and greater than

If the electrons emit a light wave which is 270° out of phase with the light wave shaking them, it will cause the total light wave to travel more quickly. This is called "anomalous refraction", and is observed close to absorption lines, with x-rays, and in some microwave systems. It corresponds to a refractive index less than 1. (Even though the phase velocity of light is greater than the speed of light in vacuum c, the signal velocity is not, as discussed above). If the response is sufficiently strong and out-of-phase, the result is negative refractive index discussed below.[15]

If the electrons emit a light wave which is 180° out of phase with the light wave shaking them, it will destructively interfere with the original light to reduce the total light intensity. This is light absorption in opaque materials and corresponds to an imaginary refractive index.

If the electrons emit a light wave which is in phase with the light wave shaking them, it will amplify the light wave. This is rare, but occurs in lasers due to stimulated emission. It corresponds to an imaginary index of refraction, with the opposite sign as absorption.

For most materials at visible-light frequencies, the phase is somewhere between 90° and 180°, corresponding to a combination of both refraction and absorption." (8)

The complexity of this explanation is a measure of the complexity of thought necessary to rationalize the standard model with phenomena that can be explained far more clearly in a simpler model. In the so-called "standard model "electrons" are busily going about "emitting" light waves, which supposedly are actually photons behaving in a wavelike manner. Such erratic behavior is required if you insist on a particulate model of reality. It's not really necessary.

In *the simple universe*, all manifestations that have been traditionally classified as "matter" are in reality complex assemblages of energy, made of complex waveforms held coherently stable by their high energy density. A beam of light striking an "object" then consists of one set of waves of energy encountering another. Their interactions are those of constructive and destructive interference depending on their relative energy levels and their range and mix of frequencies and wavelengths. So a transparent object can be seen as having a low resistance to the wavelengths and frequencies of that range of frequencies we identify as light. In its passage, the light finds few destructive interactions and probably few constructive ones as well, and is thereby only marginally changed in its passage. In transparent substances that index is likely to be a measure of the presence of energy at frequencies closely corresponding to those of the transmitted light and their effects of reinforcing and destructive interference. Visually transparent glass. for instance, is often resistant to the transmission of light in the ultraviolet range. A translucent material, on the other hand exerts a marginally greater resistance, that is, a greater amount of destructive interference occurs in the passage of the light. The absorption of color is explained in the same way. In colored "materials" destructive interference is greater at particular wavelengths and frequencies. Opacity is then the result of high degrees of complex waves of energy (high energy density) in the absorptive "material" resulting in little or no passage of light energy. In these instances absorption of the energy of the light results in warming, electrical phenomena, or structural changes in the absorbing entity. All is explained by the interactions between two (complex) sets of waves of energy. No mysterious "duality" is needed to explain the experimental result. No "electrons" are displaced, none have their "spins" modified, none are shoved into new orbits.

5. Conductivity, resistance, insulation

The focus, orientation, direction of electrical conductivity undoubtedly follows the same laws as those of transparency to light. These are basically the rules of constructive and destructive relationships between the signal and its carrier. The significant difference is that the electrical activity is carried, conducted, within the boundaries of the conductor. Mostly true, that is. The active principle here is that the frequency and wavelength of the "current" is structurally, formatively compatible with that of the carrier. If fully compatible in its structure the energy of the current will be relatively unimpeded, but again, as in transparency, this means that there is no destructive interference in that passage. Just as an electromagnetic signal such as FM radio is carried on one of a different form and magnitude, so is the "electric current" in say, a copper wire, it is carried as a temporary deformation of the EM structure of the copper. In other words, a copper wire carrying an electrical current is not structurally or compositionally the same as one not engaged in this behavior. Is this any different, you say, than saying that the copper is progressively ionized during the current-carrying activity? Of course not. The difference is in your mental model of the process, your recognition of the wave-like structure of both participants in this phenomenon and the resultant simplification and clarity of understanding that results.

Similarly current passing through a resistive medium is like that of translucency to light frequencies. Resistance results from a partial compatibility between the parties to the transaction. The result, as with light is absorption and conversion of parts of a portion of the energy from one state to another, resulting in heat, possible structural changes, or destruction of one or the other of the materials.

We said earlier that the current is carried "mostly" within the physical confines of the carrier. But take the example of long range high voltage lines. We know that there is significant "leakage" of electromagnetic radiation in the vicinity of such carriers. Of course we now can see that the electromagnetic fields created in those instances are in fact simply high intensity distortions of the background field, caused by the presence of the very high energy of the "electrical transmission." These fields are present even when the transmission lines are wrapped in insulating material, so the answer is not leakage of current, but distortion of the field.

Those insulating materials are, of course, substances whose EM structure offers a high degree of destructive interference to the primary current-carrying frequencies.

6. Wrinkles in a mist

How is it that modern physicists "know" that our universe is made up, at the smallest scale, of tiny entities we call particles? How is it that a whole branch of modern knowledge is labeled "particle physics"? Is it simply because the Greeks, Democritus, Epicurus, Lucretius, told us so? Of course not. But their setting that up as our mental model led us to lean that way when the question arose. Modern chemistry led our most serious students into the development of models of structure of the differing elements that could be easily quantified, could be repeated, and predicted. When we laid all that out, it was simplest to do it in terms of particles. We saw ones called electrons as the smallest, then arrived at protons and neutrons, those with or without electrical charges. And even though none of these was large enough to be seen with our eyes or instruments, the models we conjectured worked and we could extrapolate all the way up to bushels, tons, even megatons and they still worked. But at the bottom, it was particles.

But might there be smaller ones? When someone saw elements spontaneously change, or give off energy in a strange way, something smaller than an electron seemed it might be the answer. And we began to look for those, at least for a way to detect their presence. I'm not sure how the discovery of the cloud chamber came about, but it must have been in the search for some way to detect the phenomena that had no other apparent cause, like the fogging of photographic plates in the Curies" radium studies, or mysterious effects on some other experiments. At any rate, it was found that one could set up a glass box filled with a supersaturated vapor in the presence of a radioactive material and mysterious tracks would appear in the vapor.

Now, before we go any further, imagine yourself as a primitive human, perhaps an aboriginal on walkabout, in a hilly region say, of Australia. As you are walking you hear a sound, look up, and see a stone rolling down a rocky hillside. It was already rolling as you looked up, but you can see where it began its fall, because it has left an irregular track in the sandy soil. Simple cause and effect. You've seen those before. You walk on, watching around you, and the ground shakes.

You again look up and see another "track," irregular as before. But where is the stone? Eventually you realize that the second track is, in fact, a fracture, a wrinkle in the hillside, because the shaking ground was a small earthquake and the sandy hillside has fractured.

When the first track was seen in a cloud chamber, there was no question as to what was being seen. It was a track, left by the passage of something through the medium in the box. Something too small to be seen directly but the evidence of its passage was unmistakable. Some unseen thing has passed this way! It was completely understandable. It must be a particle. And there were others, Some went mostly straight through. Some veered off as in collision with other invisible things like water molecules. And if you held a magnet to the glass, the paths of some of the particles veered that way or circled in different ways. This confirmed that there must be many kinds of particles. The hunt was on. And that is where we are today.

But go back to when you were on walkabout. Could there be another reason for mysterious tracks to appear?

Here in *the simple universe* we can see another source for what the aboriginal saw as a track. The medium in which everything exists and appears is easily disturbed. From the "track" in a cloud chamber to the jagged stroke of a lightning bolt from cloud to ground is a fracture, a visible wrinkle in its medium, made manifest by the release of energy in the visible spectrum, as well as on both sides of that range. Most of these wrinkles, of course, are unseen. Each, however, represents a unique combination of the energy of the field. It could be that there as many varieties of these as the theorists have deemed as particles, but that seems unlikely in the simple universe we have postulated. But those that have been "seen" in the cloud and bubble chambers can certainly be seen in this way.

7. Quantum Theories

We have been stuck on "particles" ever since some Persian or Greek sat on a beach and let the sand flow through his fingers and thought, "What if everything came from something like this?" How did we get stuck with this notion? The simple fact is that it's something that common sense tells us. We live in a matter-centric universe. We see objects that must be made up of things that are ultimately hard and impenetrable. It's hard to grasp otherwise, even though from high school physics on we've been told how much space there is between our constituent atoms and how "cosmic rays" and other stuff go right through us all the time. But in today's subject, *quantum theory*, it seems that everything is made up of tiny invisible particles that sometimes behave like what we're sure they must be, hard little impenetrable substances, and sometimes like something we can't see, like waves. And unfortunately, when we get down to the details, there's a lot of evidence that doesn't quite fit.

We ran into a quandary 80 years or so go that is much like that of the ancient astronomers that resulted in the invention of planetary epicycles. In trying to reconcile the idea than sometimes matter at its tiniest scale seemed to behave one way, like little hard specks, and sometimes another, like the waves in a pond, we invented another unsupportable concept. We couldn't give up on the idea of particles, so it seems that we may have invented our own epicycles, this time in the form of quantum mechanics, to explain the apparent contradictions in the wave-particle observations. But this fix of the model brought along its own counterintuitive , even contradictory, baggage.

Let's back up a little. More than to anyone else, modern

physics owes an enormous debt to Isaac Newton, who gave us insight into most things we have tested and feel we can rely on here in the zone. But one thing he gave us that came into dispute was the nature of that phenomenon we know as light. Newton thought that light was made up of tiny corpuscles of something he didn't yet know what to call but that would become known as energy. His notion was generally accepted until an interesting set of experiments by a young doctor named Andrew Young, in about 1800, who wondered if light might be more like the waves on a pond and might travel in the same way.

In what is known to this day as "Young's Experiment," he appeared to prove that the transmission of light, in fact. had much more the characteristics of waves than of corpuscles. Along with this discovery, of course, came a strengthening of the belief in another of Newton's concepts, that of a "lumeniferous ether" a medium through which light was propagated, a medium like water which carried it's waves.

Almost another hundred years passed before Michelson and Morley carried out a sophisticated experiment that appeared to prove that that Newton's ether did not exist (leaving open the possibility that some other kind of ether, one that did not resist or slow down the passage of light, might actually be out there). A few years later, along came Albert Einstein, almost as great as Newton in his contributions to our knowledge of the world, who, with others, showed us that maybe light <u>was</u> more like what Newton thought it was, made up of little packets of energy, now to be called quanta.

As Darwin said in the conclusion of "The Origin of Species", "From these simple beginnings......", arose what we now know as Quantum Theory" (or more recently, as quantum mechanics).

One of the core principals of quantum theory can be expressed like this:

1. *"Everything is made up of tiny things that sometimes behave as if they were particles and sometimes behave as if they were waves, and you can never tell, only speculate, using the laws of probability, which it will be."*

The short name for this principle is "wave-particle duality." This seems to be the common understanding of the quantum mechanics faithful.

Now, along with this statement came a few paradoxes, that is, some apparently contradictory observations that are hard to explain. Like, "How is it that a particle can behave like a wave?, or vice-versa?, How can two particles seem to occupy the same space at the same time? Why does a particle change its appearance when it passes through a slit? How can two particles seem to be tightly tied together even at a distance" These arise from some observed behaviors we don't quite understand when we search the experiments and some from the math we use to describe them. To get at these contradictions, we should, perhaps, be clear about the characteristics of those things we call particles and those things we call waves. Here are two simple definitions that we can start with, drawn from a paper on the concept of wave-particle duality by lgsims 96, (a user name of a writer whom I would like to identify, thus far without success, previously quoted in the earlier discussion of wave-particle duality):

Attributes of particles:
A particle has mass, it is localized in space. Two or more particles cannot occupy the same space at the same time. A particle can have any relative velocity from 0 to almost c (the speed of light).
Attributes of waves:
An electromagnetic (EM) wave has no mass. It is not localized; it spreads out over a large volume of space. Many waves can occupy the same space at the same time. These waves have only one relative velocity c. They have attributes of wavelength, frequency,

intensity and amplitude of the disturbance (electric charge)

Now, all the experimental evidence shows that quanta, of light and other forms of energy, sometimes behave as if they were particles and sometimes behave as if they were waves. and we can't always be sure how they will turn up. In new repetitions of Young's Experiment, it appears that even a single (how do we know this?) quantum of light behaves the same and gives equivalent results to those Young observed, much like the waves in a pond, just as they did over a hundred years ago. Does the particle change its nature and become a wave? And can it change back? It seems, what we think of as particles sometimes can occupy the same space, a phenomenon called *superposition*, something particles should not be able to do. And certain kinds of particles give the distinct impression that they can affect the behavior of other particles, sometimes at a great distance, s phenomenon the quantum physics faithful call *entanglement*.

The history of quantum mechanics involves many of the most interesting characters in the scientific world. They agreed on some things and disagreed on others. Niels Bohr, Werner Heisenberg, Gerhard Schrödinger, even Albert Einstein were involved. And some of those things were about what was real and what was not. Einstein and Bohr in fact carried on a mostly gentle debate on the subject for many years. Part of this resulted in the famous comment by Einstein, that he didn't believe that God played dice with the universe. Finally, however, a group got together and discussed the whole thing and came up with what is now called "the Copenhagen Interpretation."

Daniel J. Castellano opens his paper *Ontological Interpretation of Quantum Mechanics*, (9) this way: "Quantum mechanics, to put it gently, is not the most philosophically lucid theory in physics."

Quantum Mechanics carries within it a number of contradictions, at least contradictions of two of the accepted principles of traditional philosophy as we know it. Castellano sees
these principles as, "a commonsensical binary notion of existence and a mechanistic notion of causality," that is, that a
thing cannot at the same time exist and not exist, and that an
event that occurs as a direct result of a particular action was
probably caused by that action and not by some other occurrence, both of which principles break down in quantum mechanics.

Castellano goes on to point out that Quantum Mechanics is
itself a mathematicized theory of physics, based almost completely on linear algebra, itself based on familiar, Aristotelian
principles of logic, including non-contradiction and the law of
the excluded middle, both of which it goes on to contradict in
its accepted expression. In addition, the idea that nothing in
quantum physics is real unless it is observed and that nothing
can exist unless except in probabilistic terms is almost a mir-
ror of the (usually misinterpreted) understanding of the writings of Bishop Berkeley that nothing is real except in our perception of it. Again from Castellano, "What is really going on
in nature when we are not looking? To suppose that such a
question has a definite answer entails a belief in objective reality, that is, a reality outside of the thinking, perceiving subject.
If we ask in what sense anything may exist or be when we are
not looking, we are making an ontological inquiry, for we are
not considering the attributes of this or that entity, but of the
act or reality of being as such. When we ask broad questions
about the reality of wavefunctions, which characterize every
type of physical entity, we are really asking questions of on-
tology." The fundamental difference here from quantum mechanics, is that in that world nothing is going on when we are
not looking.

So, let us assume for a moment that the first basis for quantum mechanics, as stated in part 1, that:

1. *"Everything is made up of things that sometimes behave as if they were particles and sometimes behave as if they were waves, and you can never tell, only speculate, using the laws of probability, which it will be."*

This cannot be true, since the actual explanation of it, in its details , so to speak, is self-contradictory and cannot be accepted. What then are the alternatives, if any?

Here are two:

2. *Everything is made up of particles, only sometimes we perceive it as if it were made up of waves.*

I'm not sure anyone totally subscribes to view number 2. It might be a subset of the QM folks, but let's take a look. It appears that some of the authors of quantum mechanics, in its Copenhagen Interpretation found that a property they labeled the "wavefunction" was a mathematical term necessary to explain some behaviors of particles and quanta, particularly their behavior in modern iterations of Young's Experiment. But there is some disagreement whether the wavefunction is real or just a mathematical term. Some physicists believe that the double slit experiment proves that particles can exhibit wave behavior while others believe this manifestation to be illusory and that the particle remains a discrete object. In both cases however, the reality of "particles" remains unquestioned. The debate goes on.

3. *Everything is made up of waves, only sometimes we perceive it as if it were made up of particles.*

Alternative 3 is my favored view. I consider myself a realist and an empiricist in matters of science and scientific theory. But in matters of how we can know and understand the world around us, I am convinced, with Bishop Berkeley, that everything that we can *know* is, in fact, only in our heads. But

I also believe that there is an objective reality and any suggestion that the world outside of our heads is all an illusion is ridiculous sophistry. Berkeley is often, and usually, I think, misinterpreted. He did not say, for instance, that if a tree falls in a forest and no one is there, it does not make a sound. His treatise is titled," A Treatise on the Limits of Human Knowledge," not on the issue of what might be real or not. We know that a tree falling in a forest generates a complex compression wave in the atmosphere that is conducted through that atmosphere, and if a person is within perceptual distance of that wave's generation, that pressure wave will cause a sensation in a person's hearing apparatus that he or she will interpret as a sound. Both reality and perception are served, within the attenuation range of that wave. Any other interpretation is nothing more than a word game, like Zeno's Paradox and similar exercises. What quantum mechanics appears to be, with its clear separation from reality, is a similar word game but in the language of mathematics.

As John R. Platt said in 1964, "... *you can catch phenomena in a logical box or in a mathematical box. The logical box is coarse but strong. The mathematical box is fine-grained but flimsy. The mathematical box is a beautiful way of wrapping up a problem, but it will not hold the phenomena unless they have been caught in a logical box to begin with.*" Our logical box requires that reality be served. The world exists. Its constituent parts must therefore also exist.

Alternative 3 does just that. It does not serve up logical paradoxes. It does not, as far as we can see, ask us to accept contradictions of logic. What it does do, in its assumption of a space filled with an electromagnetic medium, is to fulfill the need for a medium for the transmission of electromagnetic radiation, including light. That medium's fundamental structure can be seen as the basis for the speed limit that controls the transmission of light and other radiation. Alternative 3 says that the physical structures and other phenomena that we perceive and interact with are simply manifestations of both highly organized and sometimes less organized perturbations in that medium, and these are the forms and phenomena we detect and, within the perceptual range of our visual spectrum, actually see. This whole concept is consistent with reality. In sum, everything starts with electromagnetic waves, and, *an electromagnetic wave is, in fact, a perturbation of physical space, a space that is coexistent with and filled with, a resident electromagnetic field, the electromagnetic ether.*

The things that we cannot see but can still detect by other means include, in our near "zone of middle dimensions," phenomena such as the atmosphere, energy in the form of heat, magnetic fields. And those big things out there that we can only perceive at a great distance, those stars, planets, galaxies; the fields that generate the phenomena our lovers of mystery call dark energy and dark matter, can also be seen as distortions in the field of the ether, 'the cosmos'. *Dark matter,* for instance, appears to have all the characteristics of local magnetic fields, and these are easily demonstrated on a tabletop in the way they affect other nearby objects. In this model, *dark energy* is simply the electromagnetic field that fills the entire cosmos, residing inside and outside of everything else. The parts we see as objects, large and small, are themselves organized electromagnetic perturbations and/or distortions of that vast field.

This universe, from its tiniest elements to its most massive, can then be seen as made up of another kind of *entanglement* than those of the QM proponents, entanglements made up entirely of energy.

3.3 ZMD—*the zone of middle dimensions*

"Physics floated in my head.
Tendencies to exist. Tendencies to occur.
In the zone of middle dimensions, in the realm of our daily
experience, Newtonian physics is still a useful theory. We are
solid material bodies occupying empty space.
It comforts us to believe this."
—Elizabeth Rosner "The Speed of Light" a novel

1. Introduction

THAT SENSE OF COMFORT DEPENDS on several important factors. The first is how we understand the general predictability of the world around us, our near universe. We can measure our days, the regular appearance of the sun each morning, the sequence of light and dark, the location of objects around us, and their comforting regularity. In his brilliant analysis of how the brain, more particularly, the cerebral cortex, works, Jeff Hawkins ("On Intelligence" 2004)(1), describes a "memory-prediction model of how our brain enables us to go "to and fro in the world and up and down in it". It does this by registering sensory data about our environment, processing then storing this data in the various layers of the cortex, and then recalling those memories when we repeat an experience, and predicting what sequence of events will most likely come next. We know that when an anomaly in that predicted sequence occurs, as when the knob of the front door seems to have been moved when we reach for it, we sense it almost immediately, realize that we are attempting to open a different door, and are forced into a new responsive behavior.

A second factor is that we have lived with Newton's Laws for well over 300 years, and, although there is not necessarily universal knowledge of them, they are a part of our language, familiar and therefore comforting in themselves. Language has its own limitations, of course, and can cause confusion, particularly if it is misused either by inadvertence or design, or even by logical errors in discourse. A detailed discussion of the languages of physics and cosmology important. What does a physicist actually mean when he says "particle"? What image or concept is in his mind and can only be expressed by that by that particular symbolic utterance? What are these things mathematicians call "discoveries" as in "Einstein discovered relativity"? Why isn't it just a mathematical expression he made up?

Finally, we are products of an evolutionary ecological niche limited to a very narrow range of the vast array of phenomena that we have found in the universe, and, as a result, possess direct perception and knowledge of only a tiny part of it. The essays following in this Part 3 take up the laws we think we know and on which we actually depend in both everyday life and in some of our more complex and grander endeavors such as building machines, computing devices, even in space travel, and how those "Laws" can be seen as consistent (or not) with this alternative concept of the universe, its surroundings, its beginnings, and even, perhaps, its eventual destiny.

2. Atoms, molecules, the periodic table

Sitting at the feet of Aristotle in your first chemistry class, sometime in the middle of the fourth century, B.C., you would have learned almost all you were able to know in one lesson. All of creation, with the possible exception of the dwellings of the gods high on Mount Olympus, was made up of just four elements, earth, air, fire, and water. Earth was the most diverse, including a wide range of observable forms, from, say, goat dung to gold, but its essence was to be found in all of creation, even animals and man. As were the other elements, appearing in both substance and nature.

These were not laboratory discoveries, but assumed from the philosophers' observations and deductions. Some of what we now call elements, namely gold, silver, and copper could be more or less readily found, mined and refined for use. It was not until the 17th century that a more detailed identification of other elements began and not until late in that century did Robert Boyle (2) begin to identify others and raise the question of what exactly was "an element". Boyle defined elements for the first time as substance that cannot be broken down into a simpler substance by a chemical reaction. This concept is still taught today as an introduction to chemistry before introducing the concept of subatomic particles.

New elements were identified but no system of classification was introduced until Lavoisier, in 1789, published the first elementary chemistry text, listing many identifiable elements, but including some that were just assumed to exist, like light, and caloric, representing heat, but considered a substance. Attempts to classify the known elements, first into groups with similar qualities, then finally by the ways in which they interacted. The first full table of the then known elements was produced by Mendeleyev(3) in about 1869. Mendeleyev organized the elements and grouped them by similarity in their atomic weights, which also showed that those groups seemed to share similar chemical properties. He also predicted the discovery of new elements that would fill in the gaps in his table.

Over the next 120 or so years the table grew with the identification of new elements, gaps being filled, and with the advent of what we now call the atomic age, new ones, some with very brief lives, were actually created. With the advent of modern chemistry, and the devising of particle theory (primarily the electron), the structure of elements became a history of model building. The typical model of an atom of an element now consists of a core (nucleus) of positively charged particles (protons) and an orbiting cloud of negatively charged particles (electrons), in stable elements a balancing number. The atomic number of an element is assigned by the number of proton/electron pairs its atom contains. Variations in the atomic weight of an element is accounted for by the presence in the nucleus of additional uncharged particles (neutrons). Combining of elements into compounds became a process of transferring then sharing of electrons (chemical bonding). This description is highly simplified but the model has enabled the explanation of many complex interactions, creations of complex compounds of elements, and is a consistent predictor of other interactions between elements.

The Periodic Table of the Elements

Legend:
- Alkali metals
- Alkaline earth metals
- Transition metals
- Other metals
- Metalloids (semi-metal)
- Nonmetals
- Halogens
- Noble gases

Element name → Mercury | Atomic # → 80 | Symbol → Hg | 200.59 ← Avg. Mass

1	2	3	4	5	6	7	8	9	10	11	12	13	14	15	16	17	18
H 1 1.01																	He 2 4.00
Li 3 6.94	Be 4 9.01											B 5 10.81	C 6 12.01	N 7 14.01	O 8 16.00	F 9 19.00	Ne 10 20.18
Na 11 22.99	Mg 12 24.31											Al 13 26.98	Si 14 28.09	P 15 30.97	S 16 32.07	Cl 17 35.45	Ar 18 39.95
K 19 39.10	Ca 20 40.08	Sc 21 44.96	Ti 22 47.88	V 23 50.94	Cr 24 52.00	Mn 25 54.94	Fe 26 55.88	Co 27 58.93	Ni 28 58.69	Cu 29 63.55	Zn 30 65.39	Ga 31 69.72	Ge 32 72.61	As 33 74.92	Se 34 78.96	Br 35 79.90	Kr 36 83.80
Rb 37 85.47	Sr 38 87.62	Y 39 88.91	Zr 40 91.22	Nb 41 92.91	Mo 42 95.94	Tc 43 (98)	Ru 44 101.07	Rh 45 102.91	Pd 46 106.42	Ag 47 107.87	Cd 48 112.41	In 49 114.82	Sn 50 118.71	Sb 51 121.76	Te 52 127.60	I 53 126.90	Xe 54 131.29
Cs 55 132.91	Ba 56 137.33	57-70 *	Lu 71 174.97	Hf 72 178.49	Ta 73 180.95	W 74 183.84	Re 75 186.21	Os 76 190.23	Ir 77 192.22	Pt 78 195.08	Au 79 196.97	Hg 80 200.59	Tl 81 204.38	Pb 82 207.20	Bi 83 208.98	Po 84 (209)	At 85 (210)	Rn 86 (222)
Fr 87 (223)	Ra 88 (226)	89-102 **	Lr 103 (262)	Rf 104 (267)	Db 105 (268)	Sg 106 (271)	Bh 107 (272)	Hs 108 (270)	Mt 109 (276)	Ds 110 (281)	Rg 111 (280)	Cn 112 (285)	Uut 113 (284)	Uuq 114 (289)	Uup 115 (288)	Uuh 116 (293)	Uus 117 (294?)	Uuo 118 (294)

*lanthanides

La 57 138.91	Ce 58 140.12	Pr 59 140.91	Nd 60 144.24	Pm 61 (145)	Sm 62 150.36	Eu 63 151.97	Gd 64 157.25	Tb 65 158.93	Dy 66 162.50	Ho 67 164.93	Er 68 167.26	Tm 69 168.93	Yb 70 173.04

**actinides

Ac 89 (227)	Th 90 232.04	Pa 91 231.04	U 92 238.03	Np 93 (237)	Pu 94 (244)	Am 95 (243)	Cm 96 (247)	Bk 97 (247)	Cf 98 (251)	Es 99 (252)	Fm 100 (257)	Md 101 (258)	No 102 (259)

Why is the table of elements, now an easily recognized icon in laboratories, schools, even in advertisements for new compounds, called a "periodic" table? In 1914, Henry Moseley(4) found a correspondence between an element's x-ray wavelength and its atomic weight, He then resequenced the table based on these numbers by their nuclear charge. Earlier John_Newlands(5) in England noted that when he classified the known elements at that time into groups that there appeared an apparent periodicity — elements with similar characteristics appeared related in their atomic weights by some multiple of eight. Later studies of valence theory and chemical bonding confirmed the importance of the periodicity.

The correspondence between the periodicity of chemical elements in the Periodic Table with that of a new approach defining elements as stable state changes in the complex world of an electromagnetic ether is worthy of study. If, as Einstein asserts in his 1920 talk on the Ether and Relativity(6) that elementary particles are more likely to be just "condensations of the electromagnetic field", then their behavior, their combinatorial proclivities to form elements and compounds should be examined in that same light. As in the world of musical sound, we find reinforcements, harmonic resonances, and state changes. As van der Pol found in his experiments, stability occurs as energy levels increase but not smoothly, continuously. Between those stable levels noise intrudes where the harmonics are absent. The outcomes will still be comfortingly predictable, just as Newtonian mechanics still serves us, but the opportunities for new uses and new possibilities may be consequential. Again, adoption of the principles of *the simple universe* changes not the substance of the science but changes how we see it.

3. Mass, motion, momentum—the laws of objects

The term "objects" in this discussion is meant to include all phenomena that appear to persist in our environment that we can give unique identities to, by virtue of sight, touch, or other sensory perception. They may be as evanescent as clouds or as apparently solid as rocks, or mountains. They may be stationary or in motion. They may have many other characteristics identified by color, taste, texture, size. They can be described as having three spatial dimensions, as height, width and depth, along with their other perceptual characteristics.

Some of the phenomena we call objects are actually inaccessible to our sense of touch or measurement because of their distance from us, particularly those we identify as stars, planets, galaxies, etc. but we have devised ways of making close approximations of their size, mass, composition and distance from us.

As discussed in Part 1, many of these distant objects were once considered to be fixed in place, attached to a crystalline sphere, itself set in motion by an unmoved mover, as yet unknown. Aristotle, for instance, believed in as many as 55 of these concentric spheres rotating about the fixed earth, each carrying a set of fixed stars or wandering stars (the known planets). As our knowledge increased by both observation and conjecture, we realized that those distant objects appeared to behave in accordance with certain patterns and as the near universality of those patterns was identified, we began to call them laws.

In the same way, here on earth we observed the predictability of patterns of behavior of objects nearer to our own scale. Rocks thrown into the air always returned to earth, for example. Moving large objects along the earth required substantial amounts of effort, and when that effort ceased, so did the object's motion. And by an application again of observation, measurement and conjecture, our greatest minds realized that the behaviors of both the near objects and the distant ones might be seen to follow the same "laws".

While he was not by our current standards what we would call a cosmologist, Isaac Newton, by codifying the behavior of objects here on earth profoundly changed our understanding of the known Universe of his time by enumerating his Three Laws of Motion. The first law, that an object remains at rest unless moved by the application of some force was a simple observation that could be tested by repetitive trials.

1. Every object in a state of uniform motion tends to remain in that state of motion unless an external force is applied to it.

The second law consists of an equation that establishes a way of calculating both the force required to move an object and its effect on the object being moved

II. The relationship between an object's mass m, its acceleration a, and the applied force F is F = ma. Acceleration and force are vectors; in this law the direction of the force vector is the same as the direction of the acceleration vector. (We can also thank Newton for providing us with a sophisticated way of doing these calculations, particularly those of acceleration, by way of his invention of the differential calculus).

The third law recognizes (and universalizes) the fact that for every force applied there exists a resistance equal to the force

III. For every action there is an equal and opposite reaction.

The laws proved to be testable. They provided a way to test the strength of the unknown force that drew and held other objects to the earth, the force that we now call gravity. By further observations and measurements, Newton was able to show that the same unknown force held distant objects as large as the moon and the planets in their orbits. By extension, this force could be shown to explain how the earth's relationship to the sun and to the orbit of the moon could be used to both measure and predict the tides of its oceans.

Here in the ZMD, we have no reason to modify these laws. They have survived over three hundred years of testing and application and have formed the basis for understanding our world, technological innovation, and even application to our near universe astronomical observations and measurements. With them we can calculate and predict the tides, we can measure the orbits of natural and man-made satellites, we can send men to the moon and bring them safely back. Without them we might not even think the universe is expanding and at what velocity and acceleration.

These laws can be applied to our new vision of the structure of he cosmos with minimal modification. We can continue to apply them to events and actions of our daily lives with impunity. The only thing that might have to change is the terminology, in particular, the definitions of *object, mass, and motion*. For *object* we would substitute something like "organized field distortion"; for *mass*, we would substitute a measure of the distortion's energy level at apparent rest; a different term for *motion* is not so easy.

Thomas Edison, in the early part of the 20th century, was the first to create what was then called "moving pictures". He figured out that if a sequence of photographic or even drawn images of people or objects, depicting them in a slightly modified or "moved' position, were shown to human subjects in fairly rapid order, the illusion could be created that the object or person were in motion. This turned out be possible because each of the individual images is retained by the observing eye (and brain, as we now understand) in the senses of the observer long enough that a new image following can replace it without any sense of a gap between images, thereby giving the illusion of motion. For humans, we know that this sequence timing should be on the order of 1/24 of a second to ensure a smooth transition from image to image. Slower than this gives a jerky or interrupted sense of motion.

In more modern times, Walter Murch, the well-known film editor (*The English Patient*), in his book on film editing *In the Blink of an Eye*(4), explains the acceptance by a film viewer of sharp cuts in the action or between scenes in a movie by the fact that we almost always "blink" our eyes when changing our focus or turning our attention to a new object. Our visual physiology, then, assists our ability to suggest motion. In other words, our brains take care of the gaps, the discrepancies, and present them to us as smooth and continuous.

Another example, if we are in the baseball park and watch the path of the ball from the pitcher's hand to the catcher's glove, we assume that we are seeing the whole transition of the ball from one point to the other. Because of the way our brains receive the information from our eyes and process that information into images we recognize as smooth motion, it can be shown that we miss many points along the way. But just as in the motion picture example our brains fill in the gaps even if we know the gaps are there.

So here in the ZMD we believe that it is clear that the baseball is an object. The one that was thrown by the pitcher is exactly the same one that arrives in the catcher's glove (or in the center field bleachers if the batter is good, or lucky). This worked for Newton and it works for us.

In the universe we are postulating here, however, where "objects" are in reality simply organized distortions in the cosmic field, we must make a different assumption regarding motion. In this case the image that our eyes and brain "see" traveling at 80 to 90 mph toward the plate is a distortion that is in effect being continuously created at its leading edge and dissolved at its trailing edge so rapidly that we cannot detect the process. Remember that it only needs to happen at more than 24 times per second, a very slow velocity in the cosmic world, where speeds as fast as that of light are common. But if you are in the bleachers and catch the home run, you can still take it home and put it on your mantel. It won't disappear on you when it's standing still.

This, of course, does not explain the reports of many batters that Randy Johnson's fast ball disappears somewhere along its path to the plate only to reappear with a thunk in the catcher's mitt as the umpire calls "Strike Three!".

Of course, we know, after generations of observation, that we must also identify all of the forces involved here. The medium through which an "object" moves must be considered, as in air, or water. The resistance that we call friction; the distortion of the path by what we call gravity; the special instances of circular motion, what we know as orbits, with their opposing centrifugal forces and gravity, however those may be defined. But still the laws work, and we know how to use them. This also is a source of comfort.

4. the laws of energy

Unlike objects, our perceptions of energy are both more direct and more subtle. Our bodies' sensory mechanisms recognize nearby rises in the apparent temperature of the atmosphere or of objects around us, that is, we sense or understand that perceived heat or cold indicates an increase or decrease of energy in our surroundings. While we cannot see or taste energy we have learned to use it, to manage it, to discover new sources of it and, for the most part, to control it. As with the laws of objects, we have measured calculated and analyzed the nature and uses of energy, particularly as we have entered the industrial age and as a result, have developed what are known as the Laws of Thermodynamics. As a result we have also arrived at a sobering conclusion: that our sources of energy have a recognizable limit

These laws, unlike the laws of motion, are less easy to attribute to the mind of a single genius. They resulted from the efforts of numerous physicists, engineers and theorists, the first Law being made explicit by Rudolf Clausius (5) in 1850. Clausius also gave us the terms *conservation of energy* and *entropy*. These Laws explain the nature and workings of energy, the first part of Einstein's equation E= mc2.

Thermodynamics is the study of the inter-relation between heat, work and internal energy of a system. In simplest terms, the Laws of Thermodynamics dictate the specifics for the movement of heat and work. Basically, the First Law of Thermodynamics(1) is a statement of the conservation of energy, that in any closed system, there is a limit that cannot be exceeded, that while energy and mass might be interchangeable, there is a fixed supply of both in the universe. The *Second Law* is a statement about the direction of that conservation, that it cannot be reversed – and the *Third Law* is a statement about the impossibility of reaching Absolute Zero (0° K).

We know and understand that these laws work. We are an energy-bound species as are all others on earth (and, presumably, any other species that might inhabit any other universe in the cosmos). Try as we might we spend a enormous amount of or time, energy, and wealth in finding, corralling, transforming and consuming the potential energy stored on the earth, the external sources we can tap, as the sun, the tides, the chemical sources bound in the earth by accumulation of organic substance in gases, oil, coal, and the like. We have, some think too belatedly, begun to recognize that there is likely to be a finite limit to those efforts, and that our population growth and/or our energy consumption may also have limits.

But here in the ZMD, these laws can continue to be applied, even as their understanding raises warning flags for our future. The math works. The total energy in a system cannot be increased. In this closed system, there can be no perpetual motion machines. That's all there is. Entropy will continue to increase, period. And we may be headed for absolute zero, but not by choice.

If we look at these scenarios, our future does look hopeless. But the operative variable in this equation is in the term "closed system" The concept of the cosmos we have outlined in Book 1, however, may not be closed. If the Universe is just "really big" in Janna Levin's terms, hope may lie in that sea of energy our universe arose from. We know, that is, we believe, from our best observations, that the universe is expanding. In fact our most recent observations strongly suggest that its rate of expansion continues to increase. By all earthly measures, we should assume that such expansion must be fueled by energy from some source. There is as yet no indication that the universe is burning itself up by fueling that expansion internally, rather we continue to observe what appear the formation of new stars and galaxies. So from whence comes the energy to fuel that growth? Perhaps we are not inhabitants of a closed system after all. Of course, in *the simple universe*, absolute zero cannot be reached. It exists only as a mathematical construct, the result of extending a series of numbers, it is as unreal as the concept of infinity. We are, in fact, inhabitants of and part and parcel of a field with a temperature of about 2.7° Kelvin, not zero.

By all measures, the structures that surround us and even we ourselves can be seen as energy concentrators. We do it through a complex series of steps and systems, by ingesting food that is itself the product of complex chemical and energy-seeking processes. We can also see the possibility of other energy concentrating systems, an example of which might be a simple magnet, which appears capable of exerting force, even doing work, without any discernable energy input, loss of mass, or any other change. We use that capability in many ways, even though we lack a clear explanation for its power.

5. Special relativity

"On the Electromagnetics of Moving bodies", published by Albert Einstein in 1905, began a revolution in physical thought, long dominated by Newtonian physics. The paper opens with some matter-of-fact examples and assertions with which few could argue:

"It is known that Maxwell's electrodynamics—as usually understood at the present time—when applied to moving bodies, leads to asymmetries which do not appear to be inherent in the phenomena. Take, for example, the reciprocal electrodynamic action of a magnet and a conductor. The observable phenomenon here depends only on the relative motion of the conductor and the magnet, whereas the customary view draws a sharp distinction between the two cases in which either the one or the other of these bodies is in motion. For if the magnet is in motion and the conductor at rest, there arises in the neighbourhood of the magnet an electric field with a certain definite energy, producing a current at the places where parts of the conductor are situated. But if the magnet is stationary and the conductor in motion, no electric field arises in the neighbourhood of the magnet. In the conductor, however, we find an electromotive force, to which in itself there is no corresponding energy, but which gives rise— assuming equality of relative motion in the two cases discussed— to electric currents of the same path and intensity as those produced by the electric forces in the former case. (

Examples of this sort, together with the unsuccessful attempts to discover any motion of the earth relatively to the "light medium," suggest that the phenomena of electrodynamics as well as of mechanics possess no properties corresponding to the idea of absolute rest. They suggest rather that, as has already been

shown to the first order of small quantities, the same laws of electrodynamics and optics will be valid for all frames of reference for
which the equations of mechanics hold good.1 We will raise this
conjecture (the purport of which will hereafter be called the "Principle of Relativity") to the status of a postulate, and also introduce
another postulate, which is only apparently irreconcilable with the
former, namely, that light is always propagated in empty space
with a definite velocity c which is independent of the state of motion of the emitting body. These two postulates suffice for the at-
tainment of a simple and consistent theory of the electrodynamics of moving bodies based on Maxwell's theory for stationary
bodies. The introduction of a "luminiferous ether" will prove to be
superfluous inasmuch as the view here to be developed will not
require an "absolutely stationary space" provided with special
properties, nor assign a velocity-vector to a point of the empty
space in which electromagnetic processes take place." 6)

What follows in Einstein's paper is convincing mathematical support for all of his premises, most of which is not necessary to evaluate here, since this is not a mathematical treatise.
For our purposes it is sufficient to say that it established three
principals:

1. That the laws of motion apply equally regardless of
which reference frame in which they exist. This is essentially
the same as Galilean or Newtonian relativity, except that Einstein's theory denies the Newtonian concept of absolute space
and absolute time, and replaces them with a new continuum
called spacetime, which is proposed to be malleable and subject to distortion.

2. The velocity of light, c, (later, all electromagnetic radiation) is determined to be absolute regardless of reference
frame. (c = 299, 792, 458 m/sec.).

3. Derived from these two principles, mass and energy are
shown to be equivalent, expressed as $E = mc^2$.

Based on the establishment of c as a finite constant, that is, its propagation takes place in real time, he further derives the hypothesis that true simultaneity of two or more events cannot be determined by observation.

A further note. Einstein feels free to assert the absolute velocity of light in this theory, while at the same time denying the need for a "luminiferous ether" in which it is propagated, all without establishing any other possible causality for the constancy of c.

As has been said earlier, the mathematics may be correct, the velocity of light has been shown experimentally to be correct, and as far as can be determined, the equivalence of mass and energy has been shown to be, though not perfectly, in the correct neighborhood. There remain a few problems with the theory, however. In Einstein's formulation, both in the 1905 paper establishing special relativity, and later in the 1916 publication of general relativity, the theoretical concept of a physical spacetime is assumed but not definitively determined to be a real physical entity. "Space" in all other contexts is not a real physical entity and is at least a second order abstraction from reality as universally understood. A "quantum" of space cannot be isolated, examined, experimented with, stretched, compressed, distorted in any way. It serves a purely mathematical purpose, as a descriptor for the purpose of communicating the existence, size, shape, mass, and location of real entities. "Time" is a similar descriptor that cannot be isolated and examined, but is simply a tool by which we describe the duration or persistence of those real objects, events, and phenomena. So in order to be certain we are dealing with physical reality in our model of the real, physical world, we must somehow do away with or replace Einstein's *not real* spacetime continuum.

Similarly, his attack on the concept of simultaneity introduces a new factor in an otherwise purely physical theory. That addition is the concept of an observer. While it is surely true that two simultaneous events may be observed as occurring at different times if the observer is in a moving frame of reference relative to that in which the simultaneous events have occurred, but if the observer is conscious of his own velocity relative to the reference frame of those events, and is further aware of the constancy of the finite velocity of light, then the discrepancy of his observation with the reality of the aforementioned event can be easily calculated using simple mathematics. Now, in the arguments for the theory's assertions, it is usually assumed that the observer either hears the event (in this case he must be aware of the velocity of sound) or he sees them. But the theory's assertion is only that that the "observation" is not simultaneous, not that the events were not. The introduction of observation which cannot change the physical facts reminds one of Einstein's later criticism of quantum theories for introducing a causal observer into its equations.

In *the simple universe*, a major change in our view of space and time occurs. Neither are actually real physical entities, hence can neither be stretched, bent or distorted in any way, by the presence of some massive object, or any other means. Second, *all* is energy, so the appropriate conversion is from energy in one form, that of organized, coherent concentrations, entities we can perceive and identify, into its more diffuse form which we perceive only by its effects. The formula for this equivalence may still be true and ultimately confirmable. Third, the concept of space as a physical entity is replaced by the ether, which is physical, observable and manipulable. Fourth, that ether is, like Newton's space, absolute, and is the one true permanent, fixed frame of reference, as attested to by the one true observable constant, the velocity of light, c, which is seen as a coherent deformation in and of the ether. Internally, of course, the ether is not still. It is turbulent and energetic, else nothing would be here because the processes to create our "something rather than nothing" could not have occurred.

3.4 macrocosmos—imagining the unreachable

1.Introduction

From the time of humans' first perception of themselves as individual and collective beings capable of imagining the world, we have been fascinated by the heavens. We have dreamed of flying, after all, even as the birds and insects of our more mundane world. We have wondered what those mysterious, distant objects were, never conceding that we might never reach them, or they us. Our mythologies are full and seemingly endless, with gods and heroes who either went to the stars or came to us from them. As soon as we learned to count then calculate we gave numbers to our observations and distinguished them one from another, first by their constancy or wanderings, then by the patterns we discerned in and among them. As soon as we lived long enough to understand repeated patterns and to communicate these with our human brethren, we added time and durations to our descriptions and added another dimension to our thinking about them. We saw purpose, causality, behaviors we knew in earthly terms and extended those to our cosmic imaginings with the result that we postulated a maker or makers to the things we saw, and systems of belief, religions arose, and disputes between those who saw the world one way and those who saw it another. That outcome persists to this day.

Because of the almost unimaginable distances involved, even our modern scientists are reduced to imaginings. By "modern," I mean for the last few hundred years or so ago, from the time of Copernicus, Galileo, Newton on to our contemporaries, such as Hubble. As we have approached the present day we have had many more tools with which to measure, estimate, and compare what we see out there even as we have become more aware of the vanishing hope of actually reaching the stars. What follows are just a few of the parts of the science of the cosmos that we see as being in question and deserving of a new look.

2. general relativity

In 1916, Albert Einstein built on the revolutionary success of his theory of Special Relativity with the publication of his Theory of General Relativity, extending the reach of his thinking into the cosmos while at the same time replacing Newton's theories of gravitation with a new, mathematical and geometrical structure. As we have pointed out on several instances in the work preceding, that theory carried forward and built upon what we consider to be an untenable foundation, that "space" and "time" carried real physical characteristics. He had expressly denied in special relativity that that there was need for an ether, and this conceptual assumption was carried forward in general relativity. He did, later, concede that an ether was a necessity, asserting finally that relativity theories depended on the existence of a physical medium, first noted in a lecture at the University of Leyden in 1920. There was no hint of what it might be compose of, however, and Einstein's ether is always referred to as a "relativistic" ether. What Einstein proposed in General Relativity was a new, geometric version of spacetime, a four-dimensional continuum that out of which gravity appeared as a distortion of the continuum, not as a force at all. This distortion was likened by many explainers to that of depressions in an elastic membrane, like a rubber sheet on which massive objects lay. Any other object, large or small, when passing in the neighborhood of that massive object, would find itself drawn down into the depression, and depending on its velocity, would pass straight by, be deflected toward the mass, or if sufficiently slow, be drawn into it. These compelling distortions were not physical, but were rather four-dimensional distortions of the imaginary fabric, spacetime itself.

It seems clear from this rather simple description, that General Relativity actually describes nothing physical at all. It was just assumed that the mathematical model, described by Einstein himself in what may be an apocryphal comment, that it didn't matter what later experiments might show, was so beautiful, so elegant, that it must be true. You may recall that Plato was one of the early devotees of just such thinking. It was a new way of looking at the questions about the essence of gravity, that is. It was a new metaphor, in the language of mathematics, intended to make it more understandable, how its effects could be described in a more accessible way. The mathematics, it is true, was elegant, but in the years since Einstein's field equations first appeared, drawn as they were to link relativity to the electrodynamics of Faraday, Gauss, and Maxwell, they have shown sufficient flexibility that many researchers have devoted literally lifetime efforts to solve them in different ways so as to describe actual nature in a consistent way.

We have already pointed out many of the inconsistencies in these "standard model" theories, of which General Relativity is a pre-eminent example. First is the absence of a direct connection in its assumptions to discernable reality. It is often said that the predictions of relativity have been proven accurate and hence true, but because its stated assumptions, particularly its basis in mathematics, not the real world, are shown to be baseless, then its proven predictions must be attributable to some part of it that has been misconstrued. This is the argument we make from the position of the assumptions of *the simple universe*. If, as Einstein himself admitted, that General Relativity depends on the existence of an ether, and that ether is what he meant by spacetime, then what he called the distortion of spacetime by the presence of massive objects literally means a distortion of the ether by what we, in *the simple universe* call a distortion caused by the presence of a massive, coherent energy entity, like say, the sun. As we explained earlier, Massey's images of what he called "dark matter" (see figures 3.1 and 3.2) can easily be seen as those regions of distortion, and can also be seen as the source of the manifestations of what we have called gravity. This different vision of the distortions of spacetime can also be shown to be the origins of so-called gravitational lensing and, of course, of the bending of light around the sun, hailed in the 1920's as proof of relativity, especially if one concedes that "photons" have no mass.

3. *the big bang and the expansion of the universe*

There are many reasons for the attraction of a theory like big bang cosmology, but there are also many unanswered questions about it, not the lest of which is its "something from nothing" basis. Fred Hoyle, in his arguments against big bang theorists, said, "The reason why scientists like the "big bang" is because they are overshadowed by the Book of Genesis. It is deep within the psyche of most scientists to believe in the first page of Genesis." He may have been right, in part. For some reason or other, we also have in our psyches a deep love of explosions. Jamal Shrair, in his paper, *Current State of Cosmology and Astrophysics*, LinkedIn, 9/27/14), quotes the late Hannes Alfvén, the father of plasma physics, , as saying: "I was there when Lemaître first proposed this theory. Lemaître was, at the time, both a member of the catholic hierarchy and an accomplished scientist. He said in private that this theory was a way to reconcile science with St. Thomas Aquinas's theological dictum of "creatio Ex nihilo" – "creation out of nothing."

In his 2012 book, *How It Began,* (1) Chris Impey, while expressing reservations about the totality of the big bang, asserts that it, "sits firmly on a sturdy stool of four legs." They are:

Hubble expansion. Impey accepts that the linear relationship between redshift and receding velocity is likely true, that is, Hubble's Law.

Although this apparent finding was seriously questioned even by Hubble himself, it was adopted enthusiastically by his contemporaries and their followers. It was apparently important to have a justification for their preconceived notions and was questioned by only a few. However, Doppler effects, while clearly present in sound studies and other compression-wave phenomena, are not universally accepted as being the same in EM fields(cs).

The evolution of galaxies and quasars. The big bang predicts that high redshift galaxies will be younger than low redshift galaxies. This appears to be supported by observation, except that the "proof" is based on inconsistencies in one version of steady-state cosmology.

In any case, the growth and distances of these in a "bb" universe is based on premise 1, so the use of stellar evolution as support for big bang theories may be already compromised. However there is much doubt for other reasons. including the possible effect of intermediary entities or phenomena, gravitational redshifts, and the like.(cs)

Cosmic microwave background radiation. Impey's "convincing" evidence here is, of course, Penzias and Wilson's "discovery" of this phenomenon in 1964. Not at first linked to big bang theory, it was later taken as proof of the big bang because there seemed to be no other reasonable case for its existence even though its presence had been predicted as early as 1948.

However, this radiation was predicted before the concept of the big bang was adopted, and attributed to other causes, such as scattering from random stars, dust, etc.. If I am correct about the existence of an EM ether, this is a better and simpler explanation.(cs)

The cosmic abundance of light elements. These were thought to be released from (what?) in the initial, pre-expansion phase of the "bb" event, and consist of mostly hydrogen and helium.

This phenomenon can also be explained in a far simpler way by the cellular automata-like mechanisms probable in the methodology of creation of the simple universe.(cs)

The real problem here is that all four legs of Impey's sturdy stool rest on an unsupportable foundation, that all of *something* arose out of *nothing*. And, of course, any universe that had such a beginning must have an outer limit, reached by expanding for a time uncertain, into a field unknown.

In *the simple universe*, there is no necessity to accept either a big bang or expansion of the universe. In *the simple universe* the origin of all real entities, from the scale of the tiniest up to the most massive constructions in the cosmos is seen as the result of random interactions in the vast but turbulent either, by reinforcement, resonance, and phase transitions resulting in temporarily stable concentrations of higher energy, The distortions in the field surrounding these stable entities supports increased resonant activity and new concentrations. This kind of beginning negates the assumption of extraordinary explosive expansion and assumptions of "something from nothing" that characterize the theory of the big bang. This process of course also makes it impossible to determine a precise or even approximate age for the universe by backward progression from present-day observations. These beginning events could have occurred in multiple locations over many billions of years. There may in fact have been no "official" beginning, or there may have been many, of which only a few successfully survived and flourished. In this instance a parallel might be drawn with theories of the origin of life here on earth, but as has been pointed out earlier, only one or just a few were necessary for the observed outcomes to have been observed. If, as we have proposed, the cosmos, this electromagnetic ether, has always been present, what we designate and claim as "our" universe may consist of that collection of entities that can perceive, out to the limits of our perception, and it will only grow to the extent that we can increase our powers of observation with new technological advances. In other words, what we have seen thus far constitutes our "known" universe.

This, our portion, may, of course, be only a part of a greater entity with similar bodies, planets, stars, galaxies clusters, quasars beyond our horizonal limits. Our part may constitute a cluster, with others beyond our ken, our it may be only a part of a more general, more extensive, but still more or less random whole.

The ether we postulate is essentially limitless, and similar clusters and "universes" may populate other unseen regions. We will likely never know this for certain.

Is "our universe" then stable, expanding, or in some other sort of steady state? While current thought is that there is evidence of expansion, the redshift evidence supporting this hypothesis is still not totally accepted within the community of astronomers, and even Hubble himself had doubts as to its reliability. The apparent presence of high redshift quasars in among low redshift galaxies raises even more doubts. Redshift evidence may in fact be the result of other phenomena.

There is, of course, a powerful human need to circumscribe experience and delimit it in order to feel in control of our environment, near and far. Almost all human endeavors, even life itself, have known or knowable beginnings and endings. We actively seek this kind of closure, so acceptance of the idea of a universe with neither a start or a finish is an extremely difficult concept.

So here we are in our little piece of order, which arose spontaneously in the cosmos. from the same source and by the same mechanisms as possibly many others. As soon as our species achieved consciousness there came the question, "Where did we come from?" and we began to seek answers.

And as we saw in our lives that we as individuals arose from nowhere and after a period of life, returned to the same place, we assumed that the world around us must follow the same pattern, we began generating creation myths, then as now from little knowledge and much speculation. Our current myth is "the big bang."

The astronomer Fred Hoyle argued for an "eternal" model, called the "steady state" universe, but did not recognize any source for the new stars and universe expansion, so found it necessary to call on another untestable, unverifiable mechanism for the creation of new entities. He invented something he called the "c-field" arguing that this concept was no more unconvincing than a "big bang" creating a universe from nothing. *The simple universe* solves this problem. Evidence of the electromagnetic ether exists all around us. We, and all about us arose directly from the it, are an integral part of it, and will ultimately return to it. Others on other worlds may have done the same before us, may be doing it as we speak, or may do it long after our time.

We do not claim that this is a "steady state" model, that remains to be determined. It is not, on the other hand, a cyclical model, nothing in it suggests that. And this is not an continuously expanding model, although it could be. Expansion is yet to be proven or disproven.

The only part of this model that can be called "steady," is the ether, which, in relativistic terms, constitutes the only identifiable fixed reference-frame. That fixed reference frame is the reason that the maximum velocity of light is determined as "c." All other "constants, limits, and not-limits need to be re-examined and/or reconfirmed. I am convinced that those that are confirmed will also confirm the basics of *the simple universe.*

I strongly distrust and try to avoid "computer" analogies. They are relentlessly digital. They rely on on dualistic onn-off metaphor while the world is more complex than that. They seem to move swiftly off into information theory, which is already several orders of abstraction divorced from actual reality, but here is one that may make my premise more clear.

The ether, that is, the entire cosmos, may be seen to behave as an enormous computer, drawing its energy from the (almost) limitless field. That field, while cool in absolute terms, perhaps at about 2.75° Kelvin, is still conceivable as the boiling pot that energizes the cosmic computer which is running at a very high clock speed of about 1.65×10^{35} cycles per second, or 1/h. These cycles are local, not general, but their impact is universal. In each cycle, the rules governing energy values and relationships are invoked throughout the entire field, probably being responsible for its apparent turbulence (see the Planck satellite images) and the presence of 1/f noise (filtered out of the Planck images). The mechanisms in the field involve local reverberations, reinforcements and resonances. Everything we now perceive as ordinary matter and all other events and phenomena are the result of the application of those rules. Mysterious substances we call dark energy and dark matter can be seen simply as 1) the field itself, and 2) distortions of the field. The rules govern responses to randomly occurring energy concentrations, the relationships between those concentrations, and the creation of stable concentrations through accretion and phase transitions. These "rules," perhaps not quite as simple as those in Conway's "Game of Life," constitute the operating system of this enormous computing machine and generate the output of the system. That output is, quite simply, what we see around us, near and far.

4. origins of stars and galaxies

Of all the multitudinous mysteries of the cosmos, how we got stars in the first place is one of the most studied, imagined described; and with some of the most difficult to acquire information of any mystery. The formation of a star must have taken millions of years and we have not been watching that long. We might think we have seen some of them beginning, we might think we have seen them end, but the only thing we can tell from what we see is that some seem young and some seem old. So we speculate. We build models of how they might have begun, how they might have grown, but all is based on a series of snapshots, in cosmic time, that is. All, thus far, depend on several assumptions. First among these is that what we sometimes see, particularly when we look with our radio telescopes, is that interstellar space, that is, the empty space between stars and galaxies, carries clouds of what we assume to be gas and dust. How we now that is also not clear. Looking first rather close to home, at our own spiral galaxy, the milky way, it is estimated that that interstellar medium consists of:

".....10^{-4} to 10^6 particles per cm^3 and is typically composed of roughly 70% hydrogen by mass, with most of the remaining gas consisting of helium. This medium has been chemically enriched by trace amounts of heavier elements that were ejected from stars as they passed beyond the end of their main sequence lifetime. Higher density regions of the interstellar medium form clouds, or *diffuse nebulae*, where star formation takes place. In contrast to spirals, an elliptical galaxy loses the cold component of its interstellar medium within roughly a billion years, which hinders the galaxy from forming diffuse nebulae except through mergers with other galaxies. In the dense nebulae where stars are produced, much of the hydrogen is in the molecular (H_2) form, so these nebulae are called molecular clouds.

Observations indicate that the coldest clouds tend to form low-mass stars, observed first in the infrared inside the clouds, then in visible light at their surface when the clouds dissipate, while giant molecular clouds, which are generally warmer, produce stars of all masses." (2)(W)

A lot of assumptions. Different types of galaxies and different clouds seem to produce different types and sizes of stars, but this is the basically accepted model. It is thought that over time turbulence and distortions in these clouds, which are thought to be heavily ionized, cause clumping and gravitational forces to impinge upon the hydrogen atoms, density increases and ultimately, a star is born. As noted above, the higher the energy level of the nebula, the hydrogen cloud, the larger the star. Recently, a study by astronomers at the University of California, Berkeley, suggests that super massive stars appear to need a neighborhood of smaller stars to generate the super heating that make these colossi possible.

"In order for a rare, massive star to form inside an interstellar cloud of gas and dust, small "helper" stars about the size of the Sun must first set the stage, according to a new theory proposed by astrophysicists at the University of California, Berkeley, and Princeton University.
Massive stars between 10 and 150 times the mass of the Sun are few in number but produce the bulk of the heavy elements in a galaxy when they explode in supernovaes. Their extreme brightness makes them signposts of star formation in distant galaxies.
In a report in Nature, Mark Krumholz and Christopher McKee argue that early formation of a few low-mass stars in a cloud paves the way for later formation of a stellar big brother instead of fragmentation of the cloud into a hundred smaller clouds, which would produce only low-mass siblings.
"It's only the formation of these low-mass stars that heats up the cloud enough to cut off the fragmentation," McKee says. "It is as if the cold molecular cloud starts on the process of making low-mass stars but then, because of heating, that fragmentation is

stopped and the rest of the gas goes into one large star."
"What it comes down to is that if a cloud is cold, it tends to break up into many small pieces that become low-mass stars," adds Krumholz, who recently accepted a faculty position with the astronomy department at UC Santa Cruz. "As the cloud gets warmer, though, it can make bigger and bigger objects."" (3)

The operative process in all of this is, of course, gravity, the force that pulls it all together. Other researchers have a different take. A recent notion is that is that all of space consists of a plasma, and that interstellar space, particularly near the centers of galaxies, carries with it the highly magnetized characteristics of plasma and that it is this powerful magnetic field that creates the energy focus to birth a new star. To this particular researcher, gravity is only a special case of a magnetic field which is, in itself the result of the particular alignment of monopoles (which have never been isolated or identified) which make up the ionized particles which make up the hydrogen atoms of which stars are initially made.

But let's move in closer and imagine a star. We have to imagine it, of course, because there is no way we can get close enough to carry out an experiment. As recently as the middle of the last century, researchers were pretty sure they understood what went on inside a star. What chemical processes, what sub-nuclear workings brought on their birth and growth into the massive generators of energy they became. It went something like this. Stars are made up primarily of hydrogen, pulled together in massive quantities by gravity. Under the pressures of that force, hydrogen undergoes a transition into the next larger atom, helium, and in the process, gave off energy. Hans Bethe expressed that process like this:

$$4H^1(\text{hydrogen}) \longrightarrow He^4(\text{helium}) + 2\epsilon^+ + 2\beta + 2\gamma$$

Four hydrogen atoms (plus energy) convert to one helium atom plus 2 electrons, 2 beta particles, and 2 gamma particles. That is, the hydrogen becomes helium plus radiant energy. This was also imagined as if it were a gigantic furnace, using hydrogen as a fuel and burning it to become helium. The burning was not what we know as burning but a higher level reaction, a nuclear reaction. The life of a star was seen as finite, though very long by our earthly measures, but finite just the same. When the hydrogen fuel became nearly used up the process became unstable leading to a collapse and sometimes a secondary explosion we call a nova and supernova, themselves seen as then generating the heavier elements that make the more complex parts of the universe.

Now let's imagine it in a different way. In *the simple universe* model, those hydrogen atoms are high level concentrations of energy, organized at a certain scale perhaps analogous to what we call hydrogen. As the energy level in their nearby region increases, they become more concentrated and a phase transition process begins that causes a conversion to a higher level of complexity, analogous to what we see as helium. At these energy levels, however, the phase transition is unstable, somewhat like water at 100° centigrade. Some of the water is converted to steam. but some re-condenses into water, giving off its excess energy in the process. A star, at this stage releases its energy as electromagnetic radiation, at many frequencies, including that of visible light. In essence, a star is a high energy complex at a stage of self-organized criticality, and will continue in that state for long periods, oscillating from one phase state to another, drawing energy from the surrounding field at one level of intensity and radiating energy throughout the ether, until some other occurrence, like those that occur in cellular automata, interrupts the stability of that process and reverses or modifies it dramatically.

A simulation of the generally accepted process of star and galaxy formation has been attempted by the Millennium Simulation project at the Max Planck Institute for Astrophysics, using a giant computer system and over a month of calculations of the effects of gravity on what is first a basically uniform field of particles to demonstrate the gradual creation of a structured region of the cosmos by the effects of seemingly random interactions under the influence of Newtonian gravity alone. Its assumptions are, 1. the base field is made up of particles, 2. the space contains 20 million galaxies, and 3. galaxies are centered by massive black holes. The outcome of the simulation is described as:

"By applying sophisticated modelling techniques to the 25 Tbytes of stored output, Virgo scientists have been able to recreate evolutionary histories both for the 20 million or so galaxies which populate this enormous volume and for the supermassive black holes which occasionally power quasars at their hearts. By comparing such simulated data to large observational surveys, one can clarify the physical processes underlying the buildup of real galaxies and black holes." (4)

The results can easily be seen as similar to the process of star and galaxy formation as described in *the simple universe*. Sample images from the sequence are shown here, at two s levels of magnification:

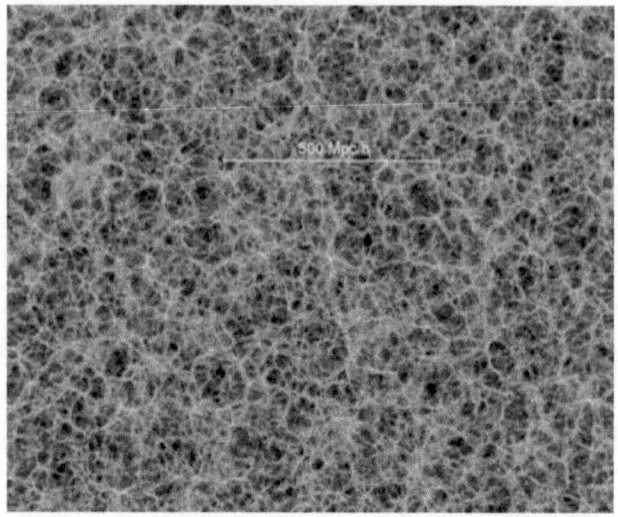

1x

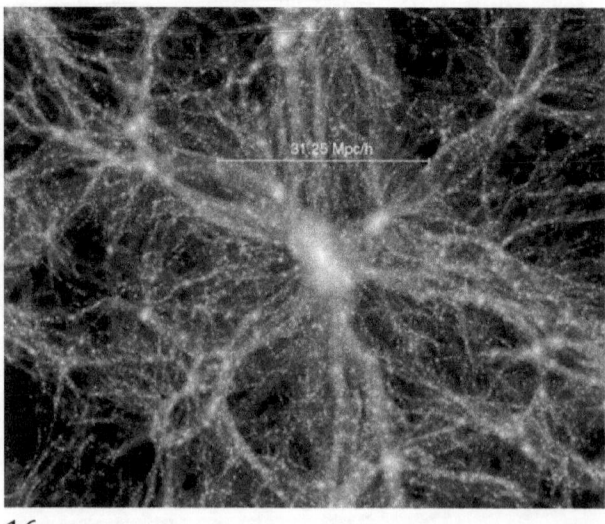

16x

This simulation process, of course, begins with the basic assumptions of big bang theory, not from its presumed beginning, but at a starting point somewhere down the line, where galaxies are assumed to be already forming. It assumes that star formation derives from gravitational attraction between particles, not from reverberations and resonances in the field. Minor deviation might be seen in the different processes, but the outcomes will likely be similar, another instance of faulty attribution. The images could be said to display either process with the same or very similar outcomes, a case of using variable parameters in formulae. Compare with an actual galaxy, M82, thought to be a star-forming region. But remember, no two galaxies are alike!

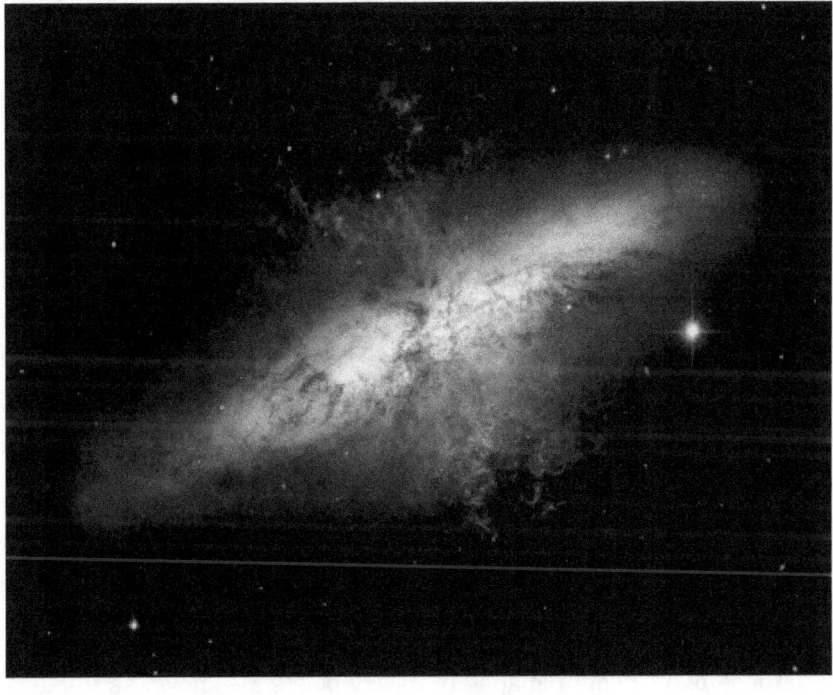

What we can take away from this speculation is agreement, somewhat, that higher energy concentrations assist in the formation of stars, that those high energy concentrations are even higher when other stars (themselves *very* high energy concentrations) are in the near neighborhood. This also takes us back to Massey's studies that show the high energy concentrations surrounding stars and galaxies (called dark matter by the establishment). High energy arises from the distortions of the field as we have seen before. The star birth theorists suggest the temperatures required are in the range of many hundreds of degrees Kelvin, while typical interstellar temperatures range from 10-20° so the presence of some significant distortions would be required, but this space is the boiling pot where everything is made and such concentrations are entirely possible. In fact, NASA's Chandra X-ray observatory, has, since its launch in 1999, mapped thousands of X-ray sources in the universe and has found many of them exhibiting temperatures in the range of 80 to 100 million degrees Celsius.

In some cases temperatures have been detected that are suspected as being too hot to permit star formation, particularly in the Perseus and Virgo regions.

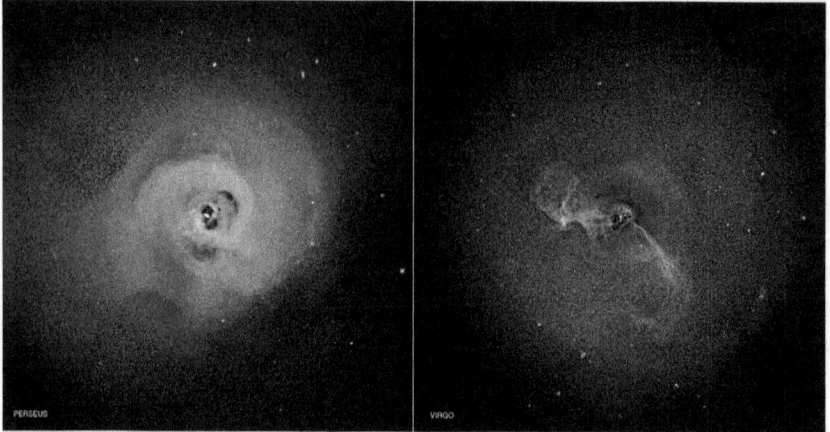

These two Chandra images of galaxy clusters - known as Perseus and Virgo - have provided direct evidence that turbulence is helping to prevent stars from forming. These new results could answer a long-standing question about how these galaxy clusters keep their enormous reservoirs of hot gas from cooling down to form stars,

"The hot (10^7 -10^8 K), X-ray-emitting intracluster medium (ICM) is the dominant baryonic constituent of clusters of galaxies. . . We find that turbulent heating is sufficient to offset radiative cooling and indeed appears to balance it locally at each radius." (5)

5. dark energy, dark matter

What are these? Well, let's fall back on Wikipedia again. In its explication of dark energy, that mysterious entity is described like this:

"In physical cosmology and astronomy, **dark energy** is an unknown form of energy which permeates all of space and tends to accelerate the expansion of the universe.[1] Dark energy is the most accepted hypothesis to explain the observations since the 1990s indicating that the universe is expanding at an accelerating rate."

Dark matter, like this:

"**Dark matter** is a kind of matter that accounts for most of the matter in the entire universe. Dark matter is one of the greatest mysteries in modern astrophysics. It cannot be seen directly with telescopes; evidently it neither emits nor absorbs light or other electromagnetic radiation at any significant level. . . ."

The existence of both is hypothesized by discrepancies between estimates of the total mass of the universe and the mass of the observable, that is, the detectable radiation-emitting matter we can "see." Since this number was determined in the 1980's the search for these mysterious substances has been unrelenting, It has been assumed that both are made up of particles, since, of course all else is. but both have been stubbornly resistant to detection. In order of magnitude, DE is assumed to make up 68.3% of the mass of the universe, DM 26.8%, and the remaining, what the ancients called "ponderable matter," just 4.9%. You can see why it is important to find it. The methods of detection have been entirely by inference, otherwise unexplainable gravitational effects and the like.

In early 2007, the journal *Nature* published the results of a study by a team of astronomers led by Richard Massey of Caltech, detailing their efforts at mapping the extent and nature of "dark matter". The university released its own notification of the publication, a portion of that press release is shown below.

New 3-D Map of Dark Matter Reveals Cosmic Scaffolding

SEATTLE—An international team of astronomers has created a comprehensive three-dimensional map that offers a first look at the weblike large-scale distribution of dark matter in the universe. Dark matter is an invisible form of matter that accounts for most of the universe's mass, but that so far has eluded direct detection, or even a definitive explanation for its makeup. The map is being unveiled today at the 209th meeting of the American Astronomical Society, and the results are being published simultaneously online by the journal Nature. According to Richard Massey, an astronomer at the California Institute of Technology who led in the map's creation, the map provides the best evidence yet that normal matter, largely in the form of galaxies, forms along the densest concentrations of dark matter. The map reveals a loose network of filaments that grew over time and which intersect in massive structures at the locations of clusters of galaxies.

Massey calls dark matter "the scaffolding inside of which stars and galaxies have been assembled over billions of years." Because the formation of the galaxies depicted stretches halfway to the beginning of the universe, the research also shows how dark matter has grown increasingly clumpy as it continues collapsing under gravity. The new maps of dark matter and galaxies will provide critical ob-

servational underpinnings to future theories for how structure formed in the evolving universe under the relentless pull of gravity.

http://media.caltech.edu/press_releases/12939

The bolded text is a clue to the underlying assumptions held in general in the astronomy and physics communities in regard to the structure of the universe, that is, that it all arrived via a "big bang" and that the dark matter scaffolding existed at a time prior to the appearance of stars, galaxies, and clusters and, in fact, provided a major part of the basis for their development. In the tradition of modern physics, "dark matter" is assumed to be particulate, that is, of the same substantial nature as so-called ordinary matter, a term that covers all of the visible, to us humans, elements of the universe.(Images from Massey's publication are shown here, and again in the chapter on gravity, following.) So, is there another possible interpretation of this impressive body of data than that proposed by Massey and his colleagues? I am convinced that there is. In his paper, *Gravitational Theory, Galaxy Rotational Curves, and Cosmology Without Dark Matter*, (6) for instance, the physicist John W. Moffat also clearly asserts, "The gravitational lensing of clusters of galaxies can be explained without exotic dark matter."

In the non-particulate universe of my own and others' concepts, "dark matter" is not "particulate," not at all mysterious, and may have little or nothing to do with expansion. It is more likely to have arisen along with the development of the large energy-dense masses of stars, galaxies, and clusters than to have been around first, to serve as "scaffolding." Dark matter is more likely to be a manifestation of the natural distortions of the field that we see around the presence of magnets or powerful electric currents right here on earth in

235

our local environment, less a "chicken *or* egg" manifestation than a more or less "chicken *and* egg" event. To explain this further, I have borrowed some of Massey's exhibits as follows:

The image below is a "dark matter" contour map of a segment of a Hubble image. According to Richard Massey, the colored points in the image are ordinary matter, i.e. stars, galaxies and clusters in the Hubble image. The contours then show the gravitational lensing intensity surrounding each of them, Massey calls these areas "dark matter," scaffolding within which new star formation is encouraged or supported. My alternative construction of this data is that the ordinary matter consists of high energy concentrations in the electromagnetic ether, concentrations we identify as stars, galaxies, and clusters, and what the contours actually indicate are the intensity of field distortions these create in their vicinity, the otherwise invisible energy distortions that, it is true, do encourage the formation of what we will ultimately see as new stars, galaxies and clusters out of the energy of the field.

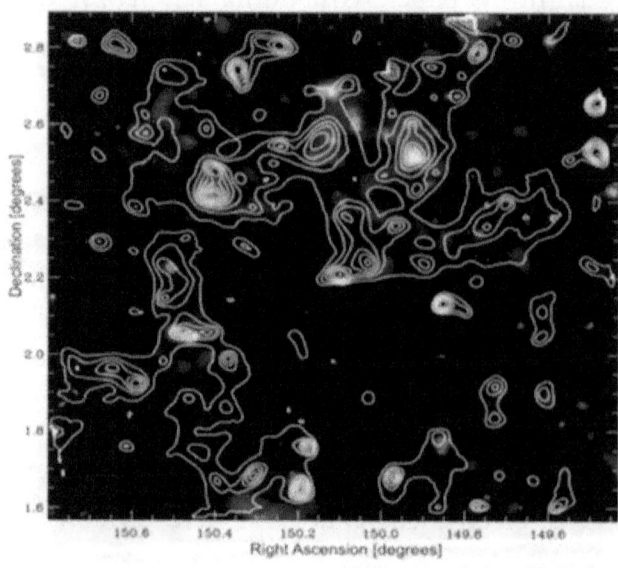

Figure 3.1 Massey's "dark matter" contour map.
(In the color image, In this one, the bright dots at the center of the various contour groups) The various colors show different types of ordinary matter but all their structure is built within the much more massive scaffolding of dark matter. (Massey: Nature)

In the computer simulation, one can visualize the location of field distortions surrounding the high energy concentrations we perceive as stars, galaxies and clusters. The apparent hard lines of the cloudlike forms are merely a representation of a selected threshold of perception. In actuality (if I may use that term) the intensity or energy level of these distortions probably follows a smooth curve, as other fields exhibit, i.e. falling off in intensity as the cube of the distance from the core. One can generalize from this conceptualization that we here in our galaxy, in fact, undoubtedly inhabit just such a region of "dark matter", but we cannot detect it as Massey has been able to do by using Hubble images of regions at a distance from us. We can also see these regions as analogous to what we do perceive locally as magnetic, electric, and electromagnetic fields, which we also know are not particulate. This can also be seen as an explanation for the apparent "curvature of space" envisioned by Einstein. Light, itself, seen now as a visible range of frequencies carried on the background frequency of the electromagnetic ether, not *through* the vacuum of space, would naturally have its path distorted in traversing these regions, thus explaining Eddington's 1919 observation of the curvature of light predicted by, and seen then as confirming Einstein's relativistic model.

In a universe that is entirely made up of a wide range of distortions of the background field, (not of little billiard balls of "particles"), these so-called regions of "dark matter" do not consist of anything like a particulate ether forcing expansion of the universe as proposed by some authorities, but are simply distortions of the field arising from the presence of the high energy distortions they surround. Ernst Mach predicted this phenomenon when he stated, "... the ether not only conditions the behavior of inert masses but is also conditioned in its state by them." (7)

So here is the truth of the mysteries of "dark energy" and "dark matter." All of everything is made up of energy, in various states depending on how and in what region of the universe it arose. That field, which we have chosen to call the ether, is all there is of what the mystery lovers among us choose to call dark energy. It is a field that while internally turbulent is fixed in its location. It is not going anywhere. Those regions of that field that have undergone distortion, increases in energy density where they surround and permeate those high energy, perceptible entities that have, up until now, been identified as objects with a property called "mass," is what those same experts have called dark matter. Our portion of the universe, our star, our galaxy, have all taken part in this process. The energy concentration we call the milky way has generated a region of distortion of the field, a region characterized by a higher energy density, a region of "dark matter" so to speak. We live in that region. It surrounds and permeates us and affects the way our stars and planets behave, as we affect it, as Ernst Mach pointed out so many years ago. And they are not made up of particles.

Powerful evidence of these assertions have appeared in two studies in recent years. British researchers, M. Lockwood, R. Stamper, and M.N. Wild published in NATURE (Vol. 399, 3 June 1999. Pages 437-439), a paper entitled *A Doubling of the Sun's Coronal Magnetic Field during the Last 100 Years.* (8)They point to a doubling of the intensity of the sun's field in that period. The reason for this occurrence is obvious in the model of *the simple universe.* Our entire solar system has moved into a region of increased energy density, that is, we have moved closer to the center of a system surrounding a high energy concentration in the field. One can picture this clearly if one imagines the orbits of our little group of planets around the sun. If the sun were fixed in position relative to the ether, or one fixed reference frame. then the planetary orbits would describe a series of concentric ellipses. But the sun itself is in motion, and into a region of higher energy density. One has then to imagine the orbits of our system as describing helices as the center of their rotation itself moves laterally relative to the ether and relative to other systems also in motion. (See Jamal Shrair, www.helical-universe.info) The fact remains, our solar system has moved in the last 100 years deeper into a region of higher energy density and the result is that our sun's magnetic field has doubled in strength.

So here arises the need for a new vision of the precise notion of gravity if "dark matter" is not actually subject to its "relentless pull", but is a totally different kind of manifestation of the structure of the field. It requires a new look at expansion, assuming it really exists, as probably the creation of new concentrations (stars, galaxies, clusters) arising with help from the universe's absorption of energy from the surrounding (and permeating) field of the cosmos.

6. Gravity

To the ancients, of course, the effects we now attribute to gravity, that is, the tendency of all free-falling objects to fall to the earth, were seen to be because they shared the same nature as the earth, and it was their desire to return to that earth that drew them down to it. Simple truth. They also attributed human emotions to trees and rocks.

Newton was not satisfied. He knew from Galileo's experiments and conclusions, that it was more complicated than this. He knew that there were more than four elements, that all objects appeared to obey the same rules, not just the "earthly" ones. He also knew that this attraction could be quantified by observation and measurement, and set out to do just that. And he also saw that these rules could apply to objects outside our reach but within our observational limits, like the moon, for instance. What he didn't know, and admitted to, was the source of that attraction. That, he left for later scientists to uncover. Newton just showed us how it worked, not where it came from.

After his success with special relativity in 1905, Einstein set to work on bringing gravity into the same framework. He was not alone in this effort. Hendrik Lorentz sought to explain it as part of his theories of electromagnetism. George-Louis Le Sage sought a particle-pressure solution, supported in part by Lord Kelvin. But no one could secure for gravity a place in the quantum/particle physics hierarchy of forces. A recent and thorough attempt at a new model is by John W. Moffat, currently working at the Perimeter Institute for Theoretical Physics in Waterloo, Ontario. I will not attempt to explain his entire theoretical basis here, but on this subject, he has this to say in describing attempts to bring all of physics together:

"Several other well-known physicists and mathematicians were also attempting to describe gravity as a generalization of special relativity. The aims of all these scientists, including Einstein, were to determine how fast gravity traveled and what in fact gravity *was*, and when those issues were settled, to find a gravity theory that incorporated all of the finite—as opposed to the Newtonian instantaneous and therefore infinite—speed of propagation of the gravitational force."(9)

In 1900, Lorentz proposed that gravity must propagate at c, like all other EM phenomena. Le Sage and Lord Kelvin agreed with this assertion, but for slightly differing reasons.

On the other hand, John Moffat sees gravity as a separate field, not tied to the other accepted forces in the universe but adding to them.

from *Reinventing Gravity,* p. 201

"The new concept of dark energy or vacuum energy—which was needed to explain the discovery of the accelerating expansion of the universe—has changed our understanding of the nature of matter and energy. Physicists have not yet even agreed on what to call this new energy. It is variously referred to as the vacuum energy, dark energy, or Einstein's cosmological constant. I think of it as a new field energy that has negative pressure and density. This exotic field energy permeates all of spacetime, in the interior of stars as well as in empty space. It can be thought of as a new kind of ether that exists universally in space time, and it does not violate the symmetries in gravity such as local Lorentz invariance or general covariance." (10)

He sees this "energy" ether as permeating all, but stops short of seeing it as the mother of all. As a result it fails to simplify, but only complicates our view of the makeup of the cosmos by adding yet another field, another set of particles to the growing panoply already postulated.

All of these prior scientists presumed that gravity was a force, or a pressure, and hence must move. Not true. Gravity is, in fact, a condition, a disturbance in space, in the ether, affecting every entity occupying the same place or region. The *simple* answer to this is that gravity does not *move*, gravity *is*.

Einstein's instinct was close to the truth. He saw gravity as a distortion, a curvature of the presumably smooth, imaginary entity he dubbed "spacetime." He had of course ignored the concept of an ether in Special Relativity, and chose not to return to it here. But he was willing to invent something out of whole cloth (a little like the creators of the Emperor's new clothes, perhaps) in taking two conceptual, not real, entities, space and time, and endowing them with physical attributes. Later, he had to concede that those needed the physical qualities of an ether or something like it in order to be able to bend, curve, etc., or to give rise to new entities like particles.

To Einstein, then, gravity was no longer a force, or a pressure of particles, but a distortion, a smooth one, a curve in spacetime, caused by the presence of massive objects like stars or planets. This is where the famous analogy of the stretched elastic membrane supporting heavy objects like balls came from, where passing objects would be drawn into collisions or possibly orbits by falling toward the depression in that elastic fabric. Nowhere is it explained how those depressions occur, however. What is the force that draws down those massive objects to create those depressions?

Needless to say, Einstein's elegant explanations did not still all criticism. There was still no explanation of the source of that force, only a new model of its behavior. And even Einstein was unsatisfied that his theory was still inconsistent with the behavior of very small objects. The quantum physicists were identifying different forces at the sub-nuclear level. Einstein's relativity equations didn't quite work to explain those, and his conviction that there must be a consistent explanation of all forces, pushed him into a thirty year, never completed struggle to find one. Others tried as well. Sakharov sought an "induced gravity" model, suggesting that gravity is not a fundamental force but an emergent one like those derived from hydrodynamics or continuum elasticity, Ver Linde has recently offered another "emergent" notion, based in part on entropy and information theory (11). None of these meets the fundamental requirement of testability, however, in the Popperian sense that a theory must be subject to disprovability to qualify as scientific. And all carry with them the underlying assumptions of the current standard models of physics and cosmology, those of particle theory, quantum theory, big bang cosmology. universe expansion, and the like. John Moffat, for example, poses a new emergent model, but his requires the assumption of yet a fifth fundamental force, also unobservable and untestable. And it requires the acceptance of a variant value for Newton's gravitational constant, "G" without a hint of causation for that allowance.

In *the simple universe*, we see gravity as an observed emergent phenomenon arising out of higher energy distortions in the field, the electromagnetic ether from which and in which all so-called "material" phenomena arise and reside. It is these distortions surrounding high density energy entities like galaxies, stars and planets that give rise to the phenomenon of gravity, and its effects, in controlling the orbits of some, of causing light to bend around stars, of supposedly controlling the expansion of the universe. This is what the lovers of mystery have called "dark matter." These distorted regions of the cosmos are the result of the presence of the very high energy-density entities of planets and stars. One could say that we stand with Einstein in that what we have previously referred to as gravity is not a force or an attraction at all but is in fact just a distortion of the field acting on us and other distortions of the field and perceived as a Newtonian "force." And the same rules apply no matter how small the focus is, down to the submicroscopic entities we identify as making up our chemical elements, each a focal point of energy concentration, each causing a distortion in the field surrounding it, each seen as exerting a "gravitational" effect on entities entering its "territory."

What is happening within these distortions? If one sees a star as a high energy concentration in and of the field, and the star is continuously giving off radiant energy, in the full electromagnetic range, continuous replacement of that energy is essential for the maintenance of existence of that star, just as a continuous flow of energy is required to maintain a boiling pot or a steam engine. The traditional view of this process is more like that of star as being just a finite cluster of fuel that when used up results in that fire being extinguished. If, on the other hand, that star is seen as needing an almost continuous replenishment of energy to maintain its "temporary" stability, its ultimate extinguishing must be the result of some other process. The only potential source of that energy is the field itself. So there must be a continuous influx of energy into the high energy concentrations we know as stars to maintain the energy density required to support that star's stability. The local manifestation of this combination of energy influx and field distortion may well be what we perceive as gravity.

And what of the "proof" of Einstein's theory? How to observe it? Well that's not so easy. The effects predicted by it were very tiny and the distances great. If Einstein were correct, the light from a distant star would be slightly deflected as it passes by a star. And we had one of those near at hand, our Sun. However, since the measurements were so miniscule and the Sun's brightness so great, it would be impossible to separate the light from a star passing close to the Sun, except at the time of a total solar eclipse. The test had to wait, partly because in 1916, there was a war going on. However, in the spring of 1919, two expeditions set out to locations on the path of an eclipse, one to Brazil, and a second led by a strong Einstein supporter, Sir Arthur Eddington. The results presented to the Royal Society, while not absolute, were taken to confirm to a high degree Einstein's predictions , that his model of gravity was totally correct. What exactly does that mean? Newton's theory of gravity would have predicted a similar result, but Einstein's seemed more precise. The question that we ask today is whether any other model could predict the same result. We think there is. Here is one recent example.

Both Eddington's team and the one sent to Brazil, suffered inclement weather that affected their observations. Today we have the Hubble Space Telescope, located far outside the effects of earth's weather systems and taking thousands of photographs with high resolution and high precision. In the course of computer analysis of thousands of those photographs one researcher, Richard Massey at CalTech, developed what he called the outlines of dark matter surrounding galaxies in the universe. We, of course, see these as distortions in the ether. Studying thousands of Hubble images, Massey was able to develop what are essentially contour maps of the cosmos showing differences in gravitational influence surrounding stars, galaxies and clusters and plotting those much as a meteorologist maps high and low pressure weather systems. One of these is shown in Figure 3.1 in the preceding section.

Massey's description suggests "dark matter" as something he calls "scaffolding" as a birthing region for new stars. As energy-dense distortions in the field these can easily be seen to have the same effect. He then generated three-dimensional computer simulations to illustrate the extent of the impact zones of these observed zones of influence. In the image below we have added examples of how these regions of distortion probably affect light in its path through and past them. The lighter line assumes no influence. The darker line shows the probable actual path of electromagnetic energy through such a region. Remember that the so-called "dark energy" regions are, in fact simply higher energy distortions of the field, generated by the extremely high energy concentrations of the stars and galaxies. The regions of greatest distortion of the field are misattributed by the standard model as being the mysterious substance "dark matter". " Gravitational lensing occurs here just as ordinary lensing results from distortion of light in other transparent media.

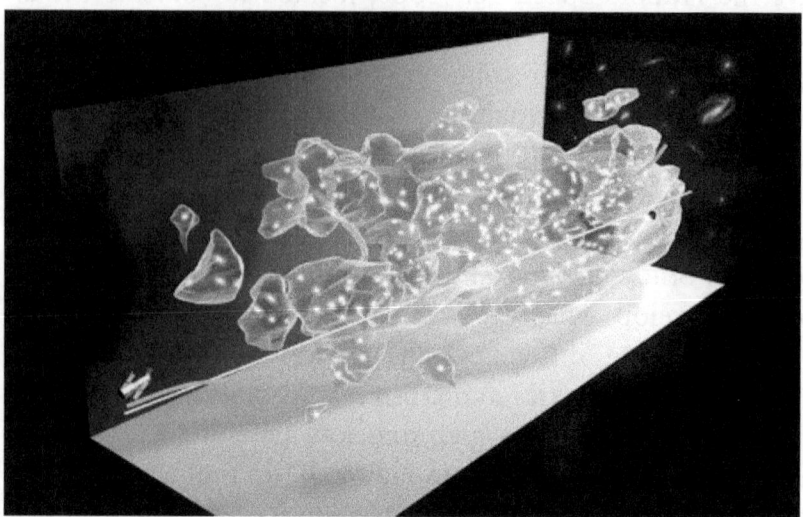

Figure 3.2 Computer simulation, "dark matter scaffolding" by Massey. *This image, created by Richard Massey through his analysis of Hubble images, gives us an idea of how the "dark matter" surrounding the high energy phenomena in the universe might be seen if it were visible. The bright elements within these distorted fields are not from the original images but were added to show how the structures might be organized. For example, our galaxy might be one of those bright dots. The straight line represents an undeflected path of light, the curved one how a light beam is likely deflected by the higher intensity of the distorted field.*

If, for a moment, we step aside from the purely mathematical speculations of Einstein, let's compare his and Newton's probable interpretations of the 1919 test. In Newton's gravity, two masses are attracted proportionately to the square of their masses. If, as modern physicists assume, light photons are massless, then no attraction (deflection) should occur. In Einstein's view, there is no physical attraction, no force, only a curvature in spacetime creating a new (geodesic?) path for light to follow.

A recent question and answer published on a physics discussion website illustrates the common view of this same quandary. The question was:

"We have seen from observations of light coming from behind objects of high mass, that the light is 'lensed' by the gravitational field of massive objects. However, light itself has no mass, so how is it affected by the gravity of these objects?"

The answer, by Sophie Allan, of the National Space Center, tries to deal with the inner paradox of the standard model in this way,

"The first point to make is that while photons (little packets of light energy) do not have mass, they do have momentum, and a change in momentum yields a force, so in actual fact light is able to physically interact with matter. However, the key to this question came when Einstein developed his theory of general relativity. Photons of light are not technically affected by large gravitational fields; instead space and time become distorted around incredibly massive objects and the light simply follows this distorted curvature of space."

If one sees, as we do, that Einstein's spacetime is, in fact, an electromagnetic field, then it becomes easy to see that the apparent gravitational lensing, the bending of the light, is the result of field distortions, not the deformation of an insubstantial entity, a mathematician's creation that has no physical characteristics.

Momentum, as well, is classically defined as mass times velocity, so to ascribe momentum to a photon is to (circularly) give it mass. The mathematicians' workaround goes like this. In the equation $p = mv$, p is the mathematical proportionality equivalent of mass, making it (mathematically) possible to attribute mass to an otherwise massless entity. Again, an abstract concept takes the place of reality in relativity/quantum world.

In *the simple universe*, both the light and the high energy entities it passes are comprised of coherent distortions of the field. We commonly accept that interacting electromagnetic phenomena have powerful distortional effects on each other. We see examples of this kind of deviation frequently in modern day physics, such as in the bending of radio waves in the atmosphere and the deflection of the solar wind by earth's magnetic field. Which interpretation of the bending of light is the more logical, which is more supported by direct experience

7. magnetism

What then, is magnetism? Is it just a variation of gravity or something totally unique? For the most part, magnetism is one of those phenomena that most believe we understand, because we know its effects, we use those effects in many practical ways, from electric motors to data storage, engines, control systems, but we really have had no clear understanding of its origins, its sources, or its causes. If one holds two permanent magnets near one another, one can detect evidence of a field effect between them. Moving one causes the other to move, but the field is invisible, insensible to touch or temperature. The atmosphere between the magnets is unchanged. Even the experts find it difficult to give clear explanations

The definitions of magnetism and magnetic fields in encyclopedias are typically unclear and when closely examined, are circular in nature. "Magnetic fields occur in the region of magnets." "Magnets are made up of magnetic particles." are the ultimate sense of these. They basically just describe the effects of magnetism. Here are two from Wikipedia.

Magnetism is a class of physical phenomenon that includes forces exerted by magnets on other magnets. It has its origin in electric currents and the fundamental magnetic moments of elementary particles. These give rise to a magnetic field that acts on other currents and moments. All materials are influenced to some extent by a magnetic field. The strongest effect is on permanent magnets, which have persistent magnetic moments caused by ferromagnetism. Most materials do not have permanent moments. (12)

A **magnetic field** is the magnetic influence of electric currents and magnetic materials. The magnetic field at any given point is specified by both a *direction* and a *magnitude* (or strength); as such it is a vector field. The term is used for two distinct but closely related fields denoted by the symbols **B** and **H**, which are measured in units of tesla and amp per meter respectively in the SI. **B** is most commonly defined in terms of the Lorentz force it exerts on moving electric charges.

In everyday life, magnetic fields are most often encountered as an invisible force created by permanent magnets which pull on ferromagnetic materials such as iron, cobalt or nickel and attract or repel other magnets. Magnetic fields are very widely used throughout modern technology, particularly in electrical engineering and electromechanics. The Earth produces its own magnetic field, which is important in navigation, and it guards Earth's atmosphere from solar wind. Rotating magnetic fields are used in both electric motors and generators. Magnetic forces give information about the charge carriers in a material through the Hall effect. The interaction of magnetic fields in electric devices such as transformers is studied in the discipline of magnetic circuits. (13)

In *the simple universe* magnetic fields and the magnetic properties of so-called permanent magnets can be seen as special cases of distortions of the ether, our universal electromagnetic field. How then do we explain the special character of magnets and their fields here on earth, that is, the existence of polarity? Well, Earth has a north pole, and we say that it also has a north magnetic pole as well. Other planets, even the sun, have polarity. Where does this phenomenon originate? Let's look at the solar system.

From a long history of observation, we know that Mercury, Earth, and the giant planets, Jupiter and Saturn, have magnetic fields strong enough to deflect that other electromagnetic effect, the solar wind, around the planet. Our magnetic field is critical to our survival since it deflects almost completely the radiation from the sun and other sources that could be fatal to most terrestrial life. The solar wind is only detectable in the upper atmosphere in polar regions as the *aurora borealis*.

A few facts about the other planets to show the diversity: Mercury, whose core is not fluid and which rotates slowly, has a magnetic field that is weak, about 1 percent of the strength of Earth's. The gas giants, Jupiter and Saturn, have very powerful magnetic fields and thus create intense zones of trapped radiation around themselves. Jupiter's interior is hugely turbulent and its magnetic field is 20,000 times stronger than Earth's. It also emits bursts of energy in the form of radio waves and "charged particles." On the other hand, Venus and Mars have almost no magnetic fields and the solar wind strikes their upper atmospheres directly. Mars' probably mostly inert interior, and Venus' slow rotation are probably why they have little or no detectable global magnetic field. What gives rise to a planet's magnetic field seems to be the presence of a layer made of a conductive and fluid material.

Eddies and currents in in that layer, like Earth's semi-molten ferrous core are thought to be the prime generator of our planet's magnetic field. Earth's moon has no detectable magnetic field. Of other moons in our solar system, only Jupiter's giant Ganymede has a detectable EM field.

Two characteristics stand out. Those objects that exhibit high energy magnetic fields appear to have a turbulent internal structure and are also in motion, rotating rapidly, both around a central axis as well as in orbit around the Sun with its own powerful magnetic field. . Earth, with its massive molten ferrous core and the gas giants like Jupiter are turbulent as well. Those with slow rotational velocity, like Mercury, and those like Mars without a turbulent core are characterized by weaker or non-existent magnetic fields. What we are led to believe is that magnetism, like gravity, is the result of a high energy concentration to begin with, modified, much like sound is modified in the atmosphere by other motions and disturbances. At least at this scale and perhaps greater, the magnetic field is generated by the movement, currents or other motion of certain "materials" (in *the simple universe*, high energy concentrations), and the polarity of these fields appears to result from their rapid rotational motion and the resultant creation of an axis of rotation.

Of course, the small permanent magnets I carry with me exhibit polarity as well, even though I know that they are not spinning rapidly in my pocket. When spun on a smooth surface they tend to come to rest in alignment with Earth's own field, indicating by their position the location of the Earth's magnetic poles. It would appear that their local magnetic field is, perhaps, a subset of the surrounding field in which they occurred or were manufactured. These small foci of magnetic energy appear to be high energy concentrators of their own, able to do work without diminishment of size, character, or

251

strength. Can they simply be drawing energy directly from the field? Both naturally magnetic objects and manufactured ones carry polarity like that of the Earth. For convenience we label those poles north and south, as we do for the earth, but we could just as well say plus and minus as we do for electrical charge. There are still questions to be explored, however, that might shed light on these issues. For example, 1) Do permanent magnets from here on earth retain their magnetic fields *and* polarity outside of earth's magnetic field, that is, in outer space? and 2) Are there magnetic objects or substances found naturally on Mars or other planets that do not have their own planetary magnetic fields?

Here on Earth there is substantial evidence that our magnetic field has actually reversed its polarity perhaps thousands of times, the most recent some 780 million years ago. Evidence has been found in both lava flows and sedimentary locations of permanent magnetic materials that have retained that reversed polarity. The earth's magnetic pole location does not match its rotational pole. In fact it moves over time requiring regular corrections in its use as a navigational aid. This movement is thought to be related to movements of its turbulent source below the crustal mantle on which we live. A very recent article in Scientific America discusses the "imminent" possibility (perhaps in the next 200 years) of the reversal of earth's magnetic poles and along the way gives a clear explanation of its source:

"Earth's magnetic north and south poles have flip-flopped many times in our planet's history—most recently, around 780,000 years ago. Geophysicists who study the magnetic field have long thought that the poles may be getting ready to switch again, and based on new data, it might happen earlier than anyone anticipated.
The European Space Agency's satellite array dubbed "Swarm"

revealed that Earth's magnetic field is weakening 10 times faster than previously thought, decreasing in strength about 5 percent a decade rather than 5 percent a century. A weakening magnetic field may indicate an impending reversal, which scientists predict could begin in less than 2,000 years. Magnetic north itself appears to be moving toward Siberia.

Geophysicists do not yet fully understand the process of geomagnetic reversals, but they agree that our planet's field is like a dipole magnet. Earth's center consists of an inner core of solid iron and an outer core of liquid iron, a strong electrical conductor. The liquid iron in the outer core is buoyant, and as it heats near the inner core, it rises, cools off and then sinks. Earth's rotation twists this moving iron liquid and generates a self-perpetuating magnetic field with north and south poles.

Every so often the flow of liquid iron is disturbed locally and twists part of the field in the opposite direction, weakening it. What triggers these disturbances is unknown. It seems they are an inevitable consequence of a naturally chaotic system, and geophysicists observe them frequently in computer simulations. "Similar to a hurricane, you can't predict [exactly] when or where a reversal will start, even though you understand the basic physics," says Gary A. Glatzmaier, a geophysicist at the University of California, Santa Cruz. Typically the local reversal peters out after 1,000 years or so, but sometimes the twisting of the field continues to spread and eventually succeeds in reversing the polarity of the entire field. The flipping takes an average of 5,000 years; it can happen as quickly as 1,000 years or as slowly as 20,000 years." (14)

—Annie Sneed, *Scientific American*

So, two conjectures. *One*, magnetism is, like gravity, a particular manifestation of the distortion of the ether, in this case the result of the inner turbulence of the cores of those massive, high energy concentrations we call stars, planets, galaxies and the like; and *two*, magnetic polarity is an emergent property of rapidly rotating magnetic entities. Both of these are consistent with the fundamental principles of *the simple universe*.

3.5. misattribution—how we got it wrong

1. Introduction

The development of a scientific theory or model can follow many different paths. The classical one is often described as moving from the scientists' discernment of patterns in his observations, followed by development of an hypothesis, testing that hypothesis through further observations and experiments, refinement of it into a theoretical model, further testing, and in modern times putting that model out by publication so that others may test it and accept or reject it. Sometimes a new theory arises as the result of many minds seeing patterns in the same or related observations, when one or more sees that a larger, more complete pattern has emerged. And sometimes a new theory emerges from the process of studying the work of others and realizing that the same data might be interpreted in a different way. Along any of these paths, unfortunately, the possibility of error can arise for many reasons.

In his critical look at modern physics, *Escape from Reality*, the author Jim Baggott attempts to structure his arguments on a set of principles that science should be following in its search for scientific truth, and then uses those as the foundation for his following criticisms. The first of these is what he calls The Fact Principal. It goes like this:

"Our knowledge and understanding of empirical reality are founded on verifiable scientific facts derived from careful observation and experiment. But the facts themself are not theory-neutral. Observation and experiment are simply not possible without reference to a supporting theory of some kind. (1)

This is both a truth and a warning. It reminds us that our observations are, of course, not just random fact collection but are guided by our preliminary ideas of what we are looking for, a proto-theory, perhaps, so that we can be productive in our search, not just wandering willy-nilly over our landscape. And, it is a warning, that we should be careful not to let the proto-theory lead us to discard observations that might challenge or contradict it.

We have pointed to many places where we believe that modern physics has gone astray. Some through simple lack of knowledge and over dependence on theory, some through rigid adherence to weakly supported theory. Actual error, in experiment or observation is only a small part of the problem. In fact it is only through error, at times, that the way to the truth becomes known. Jacob Bronowski points out that it is the optimism of scientists, carrying them from the discovery of error to repeated new efforts, that is a prime driver of scientific discovery. Error comes about through lack of knowledge, inadequacy of available knowledge, but more often through the misinterpretation or misattribution of existing knowledge.

We gain that knowledge through careful observation and experiment, from seeing patterns in the data so acquired, and from further observations and experiments to verify our assumptions about that knowledge. The scientific method then helps us to create theories and models that, if we are lucky as well as smart, to make predictions about what we might find next. As one can see, it is not a simple, straightforward process, and there is much room for error on the way. We must be careful that our first guess at the appearance of a pattern does not subconsciously lead to see only what confirms that guess, or leave out something that doesn't confirm it, or, to become so enamored with beauty of our pattern that we leave out all conflicting data.

No one is totally free from these tendencies, particularly if one has been searching for a pattern for a long time and has only now begun to see it glimmering in the darkness. Baggott gives us an example of he dangers of depending too strongly on the theory by citing Kepler's derivation of his three laws of planetary motion from the meticulous observations of Tycho Brahe, the Danish astronomer. Brahe's data showed near conformance to the ideal of the planets' circular orbits around the sun, except for a small deviation of 8 minutes of arc in the orbit of Mars. Kepler could not accept that much deviation and his recalculation of their orbits showed them to be the elliptical form we know today. Brahe's circular orbits had, of course, nearly 2000 years of acceptance as a theory, but Kepler saw the discrepancy for what it was, and saw the need to take another look. To go back even further we have the examples of Ptolemy's epicycles, driven by the conviction that the earth must be the center of everything, so all observations were driven by that theory. In recent history there was the search for the mystery planet Vulcan, believed to exist because of variations in the orbit of Mercury, later explained by other observations. What Baggott deduces from scientific history is mostly ". . *confusion and muddle, vagueness and error, good fortune often pointing the way to right answers for the wrong reasons.*"

Modern physics has its own set of examples. As pointed out earlier, the evidence of the existence of particles is twofold. One is the mindset for over 2500 years that there exist at bottom something "uncutable." The second the evidence drawn from their what was perceived to be their "tracks.". Other explanations exist, but still, everything is attributed to the presence or, in the case of many of the theoretical ones, particles "that should exist" because that is the only way to satisfy the math. As we have pointed out, much of modern theory is based on assumptions that do not stand up well to logical analysis or even experimental confirmation. From Einstein's space that has no physical characteristics, Einstein's time that his friend Gödel showed could not be proven. Bohr's, Heisenberg's, and Schrödinger's wave-particle duality, superposition, and entanglement, contradictions that must be ignored as theory, since everything else seems to work (mostly). In all of these cases, there were observations and patterns, but when the contradictions were pointed out, it was seen as best to say that they were insignificant, or that some things just could not be known, or that some future theorist's scientific discovery will clear that up.

Unfortunately that approach has led us into a flood of speculative theories themselves unsupported by data or experiment, to the detriment of any re-examination of the data.

An example of this magical thought is the definition of one of what is considered truly fundamental in modern physics. The quark's name is drawn from James Joyce's novel, *Finnegan's Wake*. ("Three quarks for Muster Mark!"). Except for a modicum of sentence structure, the Wikipedia description below could easily be buried in that nearly incomprehensible book and thought perfectly reasonable. It could have been written by Joyce!

"A **quark** is an elementary particle and a fundamental constituent of matter. Quarks combine to form composite particles called hadrons, the most stable of which are protons and neutrons, the components of atomic nuclei.[1] Due to a phenomenon known as *color confinement*, quarks are never directly observed or found in isolation; they can be found only within hadrons, such as baryons (of which protons and neutrons are examples), and mesons.[2][3] For this reason, much of what is known about quarks has been drawn from observations of the hadrons themselves.

There are six types of quarks, known as *flavors*: up, down, strange, charm, bottom, and top.[4] Up and down quarks have the lowest masses of all quarks. The heavier quarks rapidly change into up and down quarks through a process of particle decay: the transformation from a higher mass state to a lower mass state. Because of this, up and down quarks are generally stable and the most common in the universe, whereas strange, charm, bottom, and top quarks can only be produced in high energy collisions (such as those involving cosmic rays and in particle accelerators).

Quarks have various intrinsic properties, including electric charge, mass, color charge and spin. Quarks are the only elementary particles in the Standard Model of particle physics to experience all four fundamental interactions, also known as *fundamental forces* (electromagnetism, gravitation, strong interaction, and weak interaction), as well as the only known particles whose electric charges are not integer multiples of the elementary charge. For every quark flavor there is a corresponding type of antiparticle, known as an *antiquark*, that differs from the quark only in that some of its properties have equal magnitude but opposite sign.

The quark model was independently proposed by physicists Murray Gell-Mann and George Zweig in 1964.[5] Quarks were introduced as parts of an ordering scheme for hadrons, and there was little evidence for their physical existence until deep inelastic scattering experiments at the Stanford Linear Accelerator Center in 1968.[6][7] Accelerator experiments have provided evidence for all six flavors. The top quark was the last to be discovered at Fermilab in 1995." (2)

The ways we went wrong include many of the logical and linguistic errors that spring up in disciplines other than physics as well as some that are unique to our field. Logical type conflation appears across the board, in religion, politics, in general discussion. How many time have you heard the expression, "talking past each other?". The world of physics is ripe for the creeping in of mysticism, since we are dealing with many phenomena that are almost invisible or perceivable only by their effects on other entities or phenomena, so finding out that some of the leading physicists of their time were also almost mystics themselves opens many doors for wonderment. Is mathematics real? or just a useful tool? Some of us believe that this might be the greatest source of confusion in modern physics, particularly when respected leaders in the field continue to confuse one with the other. And what about these proponents of multiple invisible dimensions? Science fiction or science fact? Is any such thing provable, or is Karl Popper spinning in his grave? And finally, the big unknown, turbulence. Is it inherent in everything, or are we just waiting for sufficient computer power to finally tame this most confusing natural phenomenon? There are others, I'm sure, that we should track down and investigate, but that will probably require another book and maybe many more years of study.

2. conflating logical types

The distinction between what is real and what is imagined or perceived is a subject that has engaged philosophers from the time when consciousness first arose. To some there is no such thing as reality—we can never actually know it—it only exists for us in our minds. For those of us who believe in an objective reality, that objects, events, phenomena are actually still there in the world when we are not perceiving them, it becomes important to define the difference between what actually exists and our perceptions of them. One of the best-known thinkers on this subject was the Irish philosopher, George Berkeley, who is also one of the most misquoted. In a work titled *A Treatise Concerning the Principles of Human Knowledge*, Berkeley laid out what he perceived as this distinction. The most common quotation attributed to him is the well-known "If a tree falls in a forest and no one is around to hear it, does it make a sound?" In fact the question as stated appears nowhere in Berkeley's treatise. No matter, it is a long discussed philosophical question to which there have been many answers. To parse this statement properly, one needs to ask "What is meant by a sound?" Is it the noise perceived by an observer as a sensation in the ear or is it the compressive wave in the atmosphere caused by the falling tree that, depending on the distance from the observer, may or may not be perceived as a sound. To make this clearer, let's bring it right into the laboratory. If you and I are sitting opposite each other at a table and I strike the table with my hand, an event occurs. That event is causally connected to the compressive disturbance of the atmosphere, perceived as a sound in both our ears. But that perception is not, in and of itself, the same thing as the event of my hand striking the table. That perception is in fact an abstraction from the event itself. If I then say to you, "Why did you strike the table?" my statement is then a reaction to my hearing the sound created by that action and event and so constitutes another level of abstraction. Which of these various events, the striking of the table, the compressive sound wave created, the perception of that sound in my brain, my comment on that sound, is the actual physical point-event in this sequence. Clearly it is the first. And when describing this sequence later, what level of abstraction am I engaged in?

Alfred Korzybski, in his book, *Science and Sanity*, (3) laid out a way clarify this kind of distinction. He called these systems "orders of abstraction" and showed how confusion results in communication when these are confused. It is often described as "talking past each other" when in a discussion one party is describing the event and the other is describing his reaction to it. Bertrand Russell in his *Principia Mathematica* (4) further explicated this as a confusion of logical types.

Modern physics has suffered from this phenomenon since its beginnings. A major principal of Einstein's Special Relativity is that of the question of the perception of simultaneity. Certainly no one believes that in the physical world, two events cannot occur at the same moment. What special relativity asserts is that an observer in a moving frame of reference cannot perceive the two events as happening at the same time. Duh? Not that the observer's action in moving *caused* the simultaneity to be changed. That still happened. Event and observation of the event , two different levels of abstraction are conflated. In quantum mechanics, the same conflation occurs in discussing how measuring an event automatically changes it. Unless the process of measurement involves direct interference in the event itself, observation is purely passive. In fairy tale literature, in romantic songs, "wishing can make it so" is a common theme. It has no place in the physical world, but it is found there in many aspects of modern theories.

A new mathematical equation describing some aspect of the physical world is an invention of a human mind, not a "discovery" of a universal truth. A formula describing an event or predicting one is not the event itself. "Something from nothing" is a play on words, not a physical event. One mathematical physicist in his own book suggested as an example that "1" follows "0" in a numerical sequence as proof that something can proceed directly from nothing. Nowhere in his book can one find a suggestion that he was joking. Mathematics is a wonderful descriptive language. It is not reality but one or even sometimes many levels of abstraction away from the realities. it is used to describe.

3. the quantum mysticists

Anyone who has tried to make logical sense of quantum mechanics since its introduction in the early 20the century has run into questions about its contradictions and logical paradoxes. One is asked to believe that an elementary particle, a photon, or electron can be seen to behave both as a particle, that is as a submicroscopic object, an uncutable as the early Greeks called them, and sometimes as a wave, as the physicist Conway described. And we are asked to believe that this change in form happens just when we decide to observe the entity in question. We are further asked to believe that two such entities can so influence each other at astronomical distance that a change in the character of one, its spin, or handedness, for example, can automatically change the other. Of course, nothing of this sort occurs here in the zone of middle dimensions where we are constrained in our actual observations to events we can actually identify. These assertions are part of the logical type confusion that occurs when mathematics constrains reality in the mind of the theorist. How did this come about? Was the beauty of the math sufficient to fog the minds of other hard-headed researchers, as physicists are thought to be? Or was there some other factor in mindset of the brilliant thinkers that we don't understand?

Some of these questions are discussed in a 2009 paper published in the European Journal of Physics by Juan Miguel Marin, titled "'Mysticism' in quantum mechanics: the forgotten controversy." Marin opens his paper with this introduction:

"A few years ago In *Science* Charles Seife discussed some variants of the hypothesis that consciousness plays a role in quantum processes. He claims 1) The idea attracted a few physicists, some consciousness researchers, and a large number of mystics. The latter were also the subject of a New York Times article.
"Half a century ago Eugene Wigner venture that consciousness was the key to this mysterious process. Wigner thereby and inadvertently launched 1000 new age dreams books like *The Tao of Physics* and *The Dancing Wu-Li Masters* have sought to connect quantum physics to eastern mysticism. "
It is widely believed that Wigner was the first to introduce the hypothesis and his 1961 paper remarks on the mind-body question. Some of those who read Wigner agree with his identification of Bohr and his colleagues as the first to introduce consciousness in quantum physics." (5)

While Bohr was praised for his efforts to preserve "objective description" to modern physics against the common but totally groundless view that embraced the entry of human consciousness into the description of atomic phenomena. And while Bohr continued in this practical manner, it is clear that his contemporaries and collaborators, among them Wolfgang Pauli and Heisenberg were still somewhat open to these notions. Pauli was heavily influenced by Schopenhauer's sympathy for eastern mysticism. Both Erwin Schrödinger and Arthur Eddington were keen students of the east.

Marin makes an important case that the paradoxes of the quantum world could have clearly derived from the minds of otherwise objective scientists who were at least open to the notions that "mind and body are one" and that the will and thought processes in general could directly affect the physical world. This is clearly a subject that needs further examination. In *the simple universe* these questions disappear.

4. *mathematics vs. reality*

As we discussed in 3.5.2, *conflating logical types*, the problem of attributing real physical characteristics to purely conceptual notions has pervaded the wrings of physicists since its beginnings in ancient societies. Our recent authors are no exception. Here are four anti-definitions to ponder, all relating to assertions we have made multiple times in this text, but worth repeating here:

1. *Space* has no substance hence no reality in and of itself. What is called space is an absence of substance. If it is designated as having limits, it can be denoted as a container or enclosure of some other substance or activity, but it has in and of itself no *physical* attributes.

2. *Time* has no substance hence no reality in and of itself. What we call time is a system of measurement used to describe the persistence or duration of objects, events, or phenomena. It exists only in units of measurement such as seconds, days, eons, conveniently derived from our observations and parsing out of regular, predictable events and durations in our world.

3. A *dimension* has no substance hence no reality in and of itself. What we call a dimension is a quality, a description, a measurement, a characteristic of an object, an event, or a phenomenon that enables us to describe or quantify characteristics of those entities and to communicate those qualities to others in terms such as microns, meters, light-years.

4. A *wave* has no substance hence no reality in and of itself. A wave is a form, a characteristic, a quality, a description of some substance. Waves can be generated and detected in fluids, as in air or water; and in electromagnetic fields, but they cannot physically exist independent of a physical medium of some kind or other.

Each of these four words are constructs we use in language to describe and communicate descriptions of real substantive objects, events, and phenomena. They are critical to our understanding of concepts used in physics, the science of real objects, events, and phenomena. But they are not in themselves real, substantive objects, events or phenomena. They are merely the words we use to describe the qualities, characteristics, forms, of real entities.

So when Einstein describes a hypothetical entity he calls spacetime, a four-dimensional substance that can be bent, curved, distorted, to induce or influence real events, objects, or phenomena, his concept is nothing more than a word game, using descriptors as substitutes for real entities, giving hypothetical substance to what are only words, hence creating a fiction substituting for reality. When a modern physicist like Lawrence Krauss postulates "*A Universe from Nothing*", the title of his 2011 book, he bases his conceptual model on hypothetical entities without substance outside of mathematical models. Other authors and theorists are equally guilty of conflating descriptions, second- or third-order abstractions, substituted for *real* point-events or objects, as we have pointed out in our discussions of Korzybski's *Orders of Abstraction*, and Russell's *Logical Types*. Of course, conceptual models have value, but they also carry a burden of having a direct relationship to the real world.

Physicists of today continue to use expressions such as "the vacuum", "the field", "space", freely, sometimes interchangeably, without qualification, to describe some entity they use in their conceptual models of the universe, or for submicroscopic physical entities like the profusion of "particles" in the various versions of what is called quantum theory. When pressed for an explanation, they often qualify their assertions with expressions like, 'well, the vacuum is not really empty', or 'empty space is filled with something', or 'it's a quantum vacuum', intended to mean that it is actually filled with something called quanta, which is another word originally invented as a descriptor, but which has been miraculously converted to an entity. And they say that perhaps these entities do not really exist, they may be "virtual" or only "probabilities" that something might exist.

What has happened here is that we have invented systems of measurement and description so that we could accurately and consistently discuss, describe, and communicate those descriptions to each other, first in words of a language, then in mathematics, another language that became a useful shorthand for words.

Mathematics is not part of the class of objects, events, and phenomena that make up the real world. Writers such as Max Tegmark, author of *The Mathematical Universe*, in which he asserts that mathematics *is* reality, are victims of their own delusions. Mathematics is, and always has been a powerful tool for description, and insofar as prediction is possible, a tool for prediction. But it is of a different logical type, a different order of abstraction than the reality it is used to describe. It may be that those whose training and experience has always been in the mathematics of physics see only that aspect of the world, but its use is too often had the effect of promoting the notion that the rough world can be made smooth.

5. how many dimensions are there?

Modern physics is full of references to "dimensions." A mathematical term in use since who knows when, it became attached with quasi-mythical associations soon after Albert Einstein coined the term "raumzeit," or spacetime in the early part of the 20th century. Actually Einstein drew the concept from two other mathematicians of his time, Poincaré and Minkowski, who first posed the use of time as a fourth dimension to clarify some of their mathematical formulae, but considered the resultant 4-dimensional construct as imaginary. Einstein instead made his "four-dimensional continuum" a central part of his Theory of General Relativity, published in 1916. For Einstein, space and time became parts of the real universe, even though neither has any real attributes that are discernable to the human sensory system.

As a young man, fascinated by science, and its literary counterpart, science fiction, I was enthralled by the notion of a mysterious fourth dimension, the idea of time travel, and possible movement into another universe through "wormholes" in spacetime, translation into a another dimension, and the like. Only later, when my education caught up with my fantasies, did I understand what was really meant by a dimension and began to question how one could be considered an actual entity in the real world. This is what I learned, and from it came a new understanding of how error can become institutionalized and almost a permanent part of our consciousness.

A dimension is nothing more or less than a descriptive term in language that refers to certain attributes of objects, events, and phenomena, the entities that make up the class of real elements in the world. As a descriptor in a language it is not real in the sense that it is discernable outside of our consciousness but occupies a level of abstraction higher than the real entities it describes. Dimensions are applied to measure, describe, communicate quantifiable attributes of objects, events, and phenomena, as in denoting the length of a line, the length and breadth of a planar object, or the length, breadth and height of a solid, what we call a 3-dimensional entity. For example, a rectilinear box can be described by stating the value of those three attributes of the object, in whatever dimensional units common to the location or region in which the object is located. Common units are, for instance, feet and inches in the U.S. and some other English language speaking countries, in meters, centimeters, etc. in those regions where the metric system is in use. But you can make up your own as long as you explain them to your audience.

For an object that has a form that is more complex that that of a rectilinear box, additional dimensions my be required to describe and measure its form and character. A toroidal form, for example, might require dimensions for its outside diameter, its cross sectional diameter, and the cross-section's rotation through 360°, in this case still only three dimensions but different ones from the previous example. You can see, however, that different types of objects require different dimensions for their appropriate description. The dimensions described are the so-called spatial dimensions. Besides being used to denote the size and form of objects, there may be dimensions that locate an entity relative to other entities, either established reference points, as used by land surveyors, or separation distances between one entity and another.

In short, I think that it is apparent that there may be an unlimited number of so-called "dimensions", but they are inseparably tied to the object, event, or phenomenon they describe; they do not exist as actual entities in and of themselves in some abstract region of the cosmos.

We use dimensions to describe events and phenomena as for instance, a storm system or hurricane. We describe it as being located so many miles from land, in a specific direction, as having winds of certain velocities at specific altitudes, as being of a certain size, as growing in intensity or weakening; each of these descriptive terms are quantifiable, and can be called dimensions.

In my fifty plus years as a practicing architect, I became intimately acquainted with the importance of dimensions. Without them, I could not communicate my desires and intentions to those who would implement my plans and turn them into useful products. I used them both to describe the products and their locations. And I used dimensions to describe other characteristics that had to be included in the final work. These included the capacity of heating and ventilating systems, the quantity of water to the plumbing systems, the voltage and current carrying capacity of electrical components. Each of these values was a "dimension."

There is another set of dimensions of course, Loosely referenced in the discussion above, in talking about air, water, electricity, and in the storm example, there is the question of movement and its characteristics of velocity and acceleration. The calculation of these elements requires another dimension which we commonly refer to as time. Time is talked about by everyone and for centuries has escaped being understood. For our discussion here, it is important to understand that time, like space, exists only as a conceptual entity in our heads, not outside of them in some mysterious realm. Like space, you cannot reach out and grasp a handful or a cup full and examine it in your laboratory. Like space, you cannot bend it, curve it, break it, slow it, or speed it up. Its place in our quiver of descriptive tools is as a measure of the duration or persistence of objects, events, or phenomena. It is also nothing more than a descriptor, its units of measure derived from the observed regular periodicity of other events such as the recurrence of day and night, or the passage of earth around the sun. We use it in this way to calculate the velocity or acceleration of objects, events or phenomena or as a measure of their duration. The old joke is correct, "Time is just one damn thing after another". It has no meaning or existence in and of itself except as it relates to the duration, the persistence of the real entities in the universe.

For the last hundred years or so, physicists have been guilty of misapprehension, misattribution, misrepresentation and downright misuse of the conceptual entities they call space and time, and what they have designated as the "dimensions" of those concepts. Einstein was not the first, only the most famous to propose a "four-dimensional" continuum (three spatial dimensions plus time) as a model for the cosmos, and to build a complex mansion of theories on that hypothetical ground. For the last thirty or so years people who call themselves theoretical physicists have built mighty dream palaces on that platform; string theory, SUSY, branes, multiverses, and the like, none of which are testable or can be the basis for predictive research or can be seen to have any relationship to the real world. I believe that this descent into unreality is the result of despair at trying to build something real on an unsupportable platform.

People ask, "How many dimensions are there?" The simple answer is, "There are as many as you need." They have nothing to do with the structure of the universe, only as a way to measure, describe, and communicate our observations. The mathematicians have great fun playing with "multidimensional" worlds and concepts. These exist only in their heads, and, it seems, leave little room for real thinking. The math is not reality, it's only one tool for describing it. But until we find a way to move physics back to its real original task, that of understanding and explaining reality, there is little hope of progress. Creation myths are of many types and varieties. Some have to do with Gods as personalities, as in the Greek and Norse mythologies. In these the world (the universe) is created as a result of a conflict or cataclysm, familial or tribal, between warring creatures in another place (or dimension), as a place to which one group or the other, typically the defeated one, is banished, or flees to. In others, worlds are the deliberate creation of a god-like creature, the Christian God for example. The purpose of this creation is left unstated. An experiment, perhaps,? a new toy? to test his powers? In Alan Lightman's satirical novel, *Mister G*, the hero wakes up one morning and out boredom, perhaps, says to himself, "I think today, I'll make a universe." Which he proceeds to do, in a delightful process which enlightens him and us. Mister G's act is of the global sort. His universe appears fully formed, even though he tinkers with it as he goes along, adding details, and eventually, inhabitants with whom he interacts in unique ways. The Christian Bible's creation myth is also global. "Let there be light!" God says, and there is light. Most scientists struggled to accept this kind of instantaneous creation, including the few who said, "We don't need a creation, the universe has always existed!"

The biblical version, a "global" type of creation, had its rebirth in the twentieth century in a new form masquerading as science. We have no record of contemporary critical reaction to the biblical model, but Georges Lemaitre's concept, of a universe expanding from an initial point, which he called "the primeval atom," drew a quick response from Hoyle, who derisively branded it as "a big bang." As often happens in our sound-bite driven public discourse, the name quickly lost its derisive connotations and became the popular name for this old, but seemingly brand new idea. The biblical one sprang, like Venus from the head of Zeus, directly from the mind of God. The big bang, on the other hand, sprang from nothing, thereby contradicting thousands of years of natural philosophy, that "something" could never emerge from nothing. Basically the argument seemed to be, "We can't hope to know what might have been there before this massive explosion, so "nothing" is as good as anything else as a possible progenitor."

the simple universe is a different kind of model. Where the accepted "standard model" is global, *the simple universe* is incremental. Where the standard model is first explosive then expansionist, *the simple universe* is evolutionary, growing by small increments, perhaps many of them over short periods of a few million years or so, but proceeding by resonances, aggregations, adaptations, trials and errors, phase transitions from criticality to higher levels of stability. Where the standard model sees the universe appearing miraculously out of nothing, *the simple universe* arises out of an entity, a field, known to exist today; and by mechanisms we can observe in our everyday *zone of middle dimensions*. Where the "standard model" is not only anti-commonsense and understandable by only a highly placed and highly favored scientific priesthood, *the simple universe* is easily comprehensible to most if not all moderately educated persons interested in the universe and its origins.

The process by which this all comes about is not so hard to understand. A simple environment and a few simple rules govern the growth and form of a coral reef, a swarm of insects, a flock of birds. Most large complex organisms, geological formations, events like earthquakes volcanic eruptions, and phenomena like hurricanes and tornados have arisen in a similar fashion. Nearly every large complex system results

273

from many small entities responding to simple rules. Some examples: a swooping flock of starlings has no leader. What we see in their formation and its fluid modulations is the result of each individual guiding its behavior by observing that of his one or two or three immediate neighbors and responding by maintaining his own direction, separation and height according to those observations. Major events like earthquakes result from the accumulation of small movements and resistances in the earths crust as it moves to a critical condition, where it takes only a small change in conditions to result in a massive change in the system. Two keys to these processes are essential to this understanding. One is the existence of an environment that within itself is in constant vibratory motion. The other is that each such environment is subject to turbulent behavior in the form of flows, currents, nonlinear conditions. The universal electromagnetic field certainly satisfies the first of these requirements. Its turbulence and the presence of pink noise (1/f noise) is clear from the need to filter those frequencies out by the Planck satellite.

These characteristics of the field support the behaviors of self-organized criticality, fractal geometry, and cellular automatism particularly resulting from its iterative vibratory nature.

the simple universe then is a product of known processes without the need for unsupportable assumptions, processes or mechanisms. It can be explained without the need for imaginary, virtual particles or contradictory, quasi-mystical inventions. It is a *quantum leap* in the direction of truth.

6. *the power of mental models*

Preconceived notions, mental models and inherited prejudices have one thing in common. They are unconscious mechanisms that color our responses to observations and get in the way of true objectivity. Physics is, of course, full of them. We have probably talked ad nauseam about the principle barrier to seeing the universe in any new way, the "particle." That one has perhaps the longest history and will take a while to get past. 2500 years of seeing the world as through a "grain of sand" perspective is pretty powerful. And even though we are now told that matter and energy are the same thing, in $E=mc^2$, we carry the mental image of tiny billiard balls or the like into every discussion of the subatomic world, even when speaking the language of mathematics. Particles have mass, even though that is now generally expressed as electrical charge; Particles have spin, like little tops, although we are told that this is not actually a physical motion, just a way of denoting some otherwise indescribable characteristic. Particles can cause point-like impressions on detectors, although this is attributed to something called a wavefunction, another inde-scribable characteristic. But they are still categorized as parti-cles, imagined as tiny dots in space.

Although Newton was not what you would call a particle physicist, he enshrined in our minds two other long-lived physical concepts. We haven't pointed to these before but here they are, *mass*, and *force*. As we just pointed out, physicists now accept the notion that, at the smallest scales, at least, the mass or weight or inertia of an object can be expressed in terms of energy, so many electron-volts. But traditional measures of mass or weight are still the norm in the ZMD. That's fine. Grams, kilograms, pounds, and tons are normal and useful, They work for us. But they do make it hard to adopt a different mode of thought, that is, to see what we have for ages considered solid, real "things" as actually being organized aggregations of energy, made temporarily stable "condensations" in a field of energy. The almost universal acceptance of the equivalence of matter and energy expressed by Einstein, has not broken through the mental model barrier to acceptance that it actually means that E and m are the same thing, not just convertible one to the other. Still stars and galaxies are referred to in the astronomical literature as "ordinary matter," and the energy-dense regions surrounding them as something more mysterious, "dark" matter.

Similarly, as Jacob Bronowski said in his Silliman lectures in 1967, "I believe that any theory that we as humans beings make at any point in time is full of provisional decodings which are to some extent as fictitious as the notion of force in Newton." *Force* has undergone some revisionism, now being generally changed to "interaction" in the modern physics literature, but the word is still used to describe the four "fundamental" forces of nature. This is perhaps a step in the right direction, but still leaves empty the class of definition of terms, which includes qualities like force, charge (positive, negative, neutral), spin, etc. To Newton, "force" was a direct translation of the metaphor of a man throwing a ball through the air which led to his imaging the moon as just such a ball, to the action of a man pushing a barrow against a resistance, which pushed back with equal opposition. No other definition of the term "force" exists, although it continues to be used in its original Newtonian sense, even in "the mechanics of quanta" to the present day.

These are just a few of the mental model barriers we work against when we attempt to "see with new eyes" in the words of Thomas Kuhn, on our way to a new scientific revolution.

7. the roughness of reality

Axiom 4. Left to its own devices nature is rough, not smooth. The Platonic ideal of perfect smoothness is not at-

tainable. Nature is irregular, not ordered. It is random, not predictable. No two snowflakes are identical, no two galaxies. Further, the universe is not homogenous, not isotropic, This "cosmological principal" is a false assumption, invented for the convenience of mathematics.

We know that Plato believed in the inherent perfection of certain forms, shapes, and lines. These perfect forms were circles, spheres, the regular tetrahedrons. Given by the gods, these were the forms that nature sought out, that unsmooth shapes tended toward, that science as he knew it was drawn to. Later scientists and thinkers, even the religious, sought platonic ideal. The cosmos was made up of crystalline spheres wherein lay the wandering stars, the fixed stars, and at the perfect center, the earth, home of man, for some, god's most perfect creation. We have read of Hippasus of Metapontum who first imagined a way to measure motion as a smooth curve of an infinite number of steps and who was sacrificed by his contemporaries for overstepping into the realm of the gods. Only when Newton created the differential calculus in a more enlightened age was smoothness, of trajectories, orbits, and the like, finally enshrined as the product of a mathematical tool. We could measure and calculate smooth motion, velocity, acceleration.

For Plato and those who followed him smoothness was an (almost) unattainable goal, a conceptual ideal. The invidious nature of that ideal was truly unknown, but it shaped the thinking of scientists for hundreds of years. What science seek is not just knowledge. What scientists are really looking for is patterns and the best of those are the ones that cross disciplines, that is that show that a s mall pattern is similar to a large one. We use this all the time scaling a large pattern down or a small pattern up and use small ones to predict the behaviors of large ones. But all of these are seen in the mindset of perfect forms. Plato's ideals live on over the centuries.

Who knows, this may be a proper approach, that a few simple understandable rules may lead to predictable behaviors at many scales, but maybe not. We have become more sophisticated. We invented new mathematics, we gave values to observations and tested them. we invented complex equations that predicted close approximations to the observed values. But nature, left to her own devices, remained rough. Nowhere in nature are the perfect ideals found, despite all of man's attempts to discover them. A difficult reality to accept. To their credit, the quantum theorists moved toward accepting this, but in the quantum world this was expressed as a willingness to accept the idea that we could never *observe* the perfect, not that it could not exist! The uncertainty principle of Heisenberg, Godel's incompleteness theorem, the emergence of "probabilities," not observations, in the physics literature are all evidence of this.

Today we can begin to see the folly of this search for the ideal. What was needed was a new mathematics of the actual, the closest to which we have today is Mandelbrot's fractal geometry. In his book *Fractals and Chaos (6)*, he lists some questions that lay out the problem with a measure of clarity. We need a geometry that answers questions such as these.

"How do we measure and compare the roughness of ordinary objects such as a broken stone, metal, glass, or a piece of rusted iron?
How long is the coast of Britain?
How can we define the speed of the wind during a storm?
What shape is a cloud, a flame, or a welding?
What is the density of galaxies in the universe?
How do we distinguish proper music from noise?
How to measure the variation of the flow of messages on the internet?

How to measure the volatity of the prices quoted on financial markets?
How to characterize the boundary between two basins of attraction in a chaotic dynamical systems?
How to characterize self-avoiding random walks?
How to characterize the critical clusters of percolation?"

We can add to this list such things as;
 How describe the shape of a moving swarm or flock?
How do we predict the frequency of earthquakes? The list can go on and on.

As Mandelbrot goes on to say:

"The word, *rough,* appears in only one of the above questions, but the underlying concept appears in *every one.* (Irregular would have been a more elegant word, but rough is more telling.) All of these questions were without geometric answer until fractal and multifractal geometry provided the beginnings of a workable and useful approach based on the surprising fact that, both in nature and culture, roughness is very often fractal."

In *the simple universe,* creation depends upon the random nature of nature. It is the very essence of how the universe as we know it came about. It is the prime characteristic of all the objects, events, and phenomena that we see about us and their relationships one to another. In the real world every surface, every shape, every motion carries this property., roughness, irregularity. It is only at the stage of phase transitions that some stability of form occurs, but in its details every form is unique. As we stated in Axiom 4, no two snowflakes, no two galaxies are the same. Similarly no two events or phenomena are exactly alike. Knowing all about initial conditions cannot guarantee predictable outcomes. Criticality is self-organizing, but not under outside control or predictability. All we can know is the underlying nature of self-referentiality. Prediction of natural events is at its best a coarse measure. The only smoothness is in the math, and that is because we made it so, having abstracted it from reality, like Newton's leaving out the moon in his unsolvable three body problem. Once again, the universe is at its best an ill-structured problem.

8. turbulence

Axiom 5. At all scales everything in the universe is in motion, both internally in all its component parts and externally in relation to the fixed frame of the ether and all other perceptible entities. This motion is essential to our perception, which functions only by the detection of differences. And turbulence is a necessary condition of that motion. Without it can occur none of the essential interactions of reverberation, reinforcement and resonance out which emerge the coherent concentrations of energy which make up the universe as we know it.

Si quaeris veritas, circumspice. "If you seek the truth (of these assertions), look about you." The very word, turbulence, comes to us from a notable source. The essence of the renaissance man, the paradigm of that title, Leonardo da Vinci, looked at the motion of water in a receptacle as it was being filled and described it both in image and in words. His word for it was "*turbolenza.*"

"... the smallest eddies are almost numberless, and large things are rotated only by large eddies and not by small ones, and small things are turned by small eddies and large."

Wile the subject of turbulence has occupied the minds of researchers for all the centuries since Leonardo. no one claims to have adequately explained it. A probably apocryphal story, attributed to both Einstein and Heisenberg, relates that the two questions they would like to ask God (in the possibility of an afterlife) would be about the truth of relativity and the solution to turbulence, in the hope that at least the first would result in an answer.

In a current course offering at the University of Kentucky, the following definition of turbulence covers the known ground succinctly:

"Turbulence is an irregular motion which in general makes its appearance in fluids, gaseous or liquid, when they flow past solid surfaces or even when neighboring streams of the same fluid flow past or over one another."
Hinze, in one of the most widely-used texts on turbulence [6], offers yet another definition:

"Turbulent fluid motion is an irregular condition of the flow in which the various quantities show a random variation with time and space coordinates, so that statistically distinct average values can be discerned."

The following is a list of physical attributes of turbulence that for the most part summarizes the preceding discussions and which are essentially always mentioned in descriptions of turbulent flow. In particular, a turbulent flow can be expected to exhibit all of the following features:

1. disorganized, chaotic, seemingly random behavior;
2. nonrepeatability (i.e., sensitivity to initial conditions);
3. extremely large range of length and time scales (but such that the smallest scales are still sufficiently large to satisfy the continuum hypothesis);
4. enhanced diffusion (mixing) and dissipation (both of which are mediated by viscosity at molecular scales);
5. three dimensionality, time dependence and rotationality (hence, potential flow cannot be turbulent because it is by definition irrotational);
6. intermittency in both space and time.

It is readily seen that none of these definitions offers any precise characterization of turbulent flow in the sense of predicting, a priori, on the basis of specific flow conditions, when turbulence will or will not occur, or what would be its extent and intensity. It seems likely that this lack of precision has at least to some extent contributed to the inability to solve the turbulence problem: if one does not know what turbulence is, or under what circumstances it occurs, it is rather unlikely that one can say much of anything about it in a quantitative sense. (7)

"In fluid dynamics, turbulence, turbulent flow is a flow regime characterized by chaotic property changes" (Wikipedia). As you can see, turbulence is considered to be one of the key unsolved problems in physics, although some interesting patterns have been discerned. For instance, "in 1933, Johann Nikuradse measured the friction experienced by a fluid when forced through a pipe at varying speeds. At first the friction gets smaller as the speed gets larger, but then at even higher speeds it increases again before attaining a constant value." We can see echoes here of our previously cited experience of van der Pols with electromagnetic outputs. Turbulence is in fact a self-organizing motion. "It is not random, tranquil water is more random...Eddies are, in some sense, the most efficient way water can flow in the given conditions...water molecules correlate their motions in order to form eddies" (like birds make flocks?)

"In another study, at the University of Illinois, physics professor Nigel Goldenfield shows that the turbulent state is indeed not random, but contains subtle statistical correlations similar to those known to exist at phase transitions, such as the onset of magnetism in crystals." Patterns on patterns emerge.

"Objects placed in a turbulent flow—even objects that are identical and which are dropped into the same spot—will end up in different places." (8) Eyink was also able to show that the magnetic lines of force carried along in a moving magnetized fluid move in a completely random way when the fluid flow is turbulent.

Turbulence is everywhere at all scales we can observe. In the atmosphere it challenges meteorologists daily in their predictions of weather. (Interestingly, in a study of disciplines engaged in predictions of various sorts, the US Weather Service was shown to rank at the top of the list., in spite of the common perception of it's unreliability.) It is apparent at a galactic scale as the image 0f galaxy M82 below, demonstrates.

the simple universe shows how turbulence in the back-ground radiation field, the ether, is the necessary precondition for the initial appearance of the energy concentrations we pos-it as the origins of the universe.

(from Tarko) ".... it seems that turbulence offers us at a macroscopic level a view of what happens during other phase transitions at microscopic level. Thus, when you are looking at eddies in a river, at turbulence in a cascade, or at cigarette smoke, you can also imagine that you are actually watching what happens to molecules during a melting process or how magnetism gradually vanishes when a magnet is melted." (9)

Or what is happening continuously in the heart of a star like our sun.

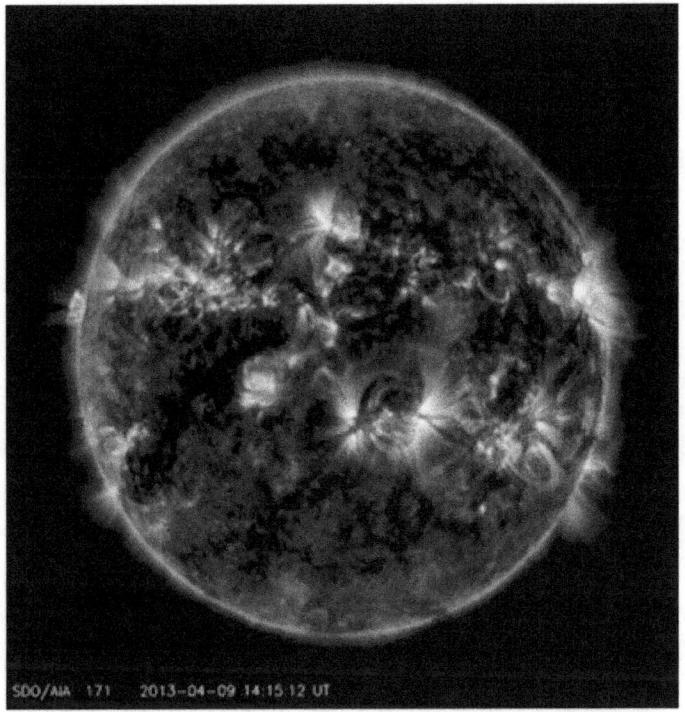

SDO/AIA 171 2013-04-09 14:15:12 UT

Or, finally, how it is that there is something rather than nothing. Turbulence is apparent in the energy of the ether, in the movements of the earthly atmosphere, in the hearts and distribution of stars, in the arrangements of galaxies, and at all levels in between. We exist in its embrace and as a result of its existence.

imagine darkness

Part 4. Afterword

This book, like the one before it, started with the observation that its inspiration was, in part, the question posed by Janna Levin in *How Did the Universe Get Its Spots?*, "Is the universe infinite, or is it just really, really big?" Infinity, of course, has no place in reality, only as a mathematical concept, what you get if you try to divide some quantity by zero, by nothing. Whole books have been written about nothing— what it is, what it isn't—but there's not much there, if you'll pardon the expression, except for the metaphysicians' endless word game. So the universe, assuming it's real, must be finite, although its actual extent is undoubtedly very difficult, probably impossible, to measure. In looking back over this text, there are three questions that have still not been adequately

answered. Here are my thoughts on those:
1. *Are there multiple universes?*

In *the picnic at the edge of the universe*, I sought to portray the universe, perhaps I should say, *our* universe, as possibly an island, a very large one, in the immeasurable vastness of the cosmic sea, but with the possibility that others might also have arisen but perhaps at a great distance away. The interludes between sections of that book depicted an imaginary voyage of discovery attempting to find one of those others.

In the school of speculative cosmology, a group I have only recently heard of, the notion of multiverse has become a popular diversion. Only today, there appeared an announcement of the publication in Physical Review X of a paper by "scientists" asserting, "*Many Interacting Worlds Theory: Scientists propose existence and interaction of parallel worlds*", this one claiming they must exist and that their interactions with ours might explain some of the paradoxes and contradictions of our current standard models. These multiverse ideas have taken many forms—scattered through the cosmos at unreachable distances from each other, located near to us but still unreachable except by travel through another dimension by way of appropritely named "wormholes" (a subject I have easily dismissed earlier).

One recent writer, Jim Holt, raises the question this way:

"Should we or should we not believe in multiple universes?. . . . . If the universe is "everything there is," then isn't it true by definition that there is only one of these things? . . .

. . . But when physicists and philosophers talk about two different regions of spacetime being "two universes," what they generally mean is that those regions are (1) very, very large; (2) causally isolated from each other; and hence, mutually unknowable by direct observation. . . .

. . . . So why believe in the multiverse.
Since other universes are, by definition, not directly observable from our own, the burden of proof is clearly on those who claim they exist." (1)

—Jim Holt, *Why Does the World Exist*? Liveright Publishing, New York, 2012

For conceptual reasons very different from those speculations, I have come to accept the idea that there may well be physical worlds well beyond our own reach, but not in any mystical sense, not in any multi-dimensional sense. We will probably never know of their existence for sure, but that is OK. To begin with, in the concept of *the simple universe* there is only one cosmos, one essentially measureless universal field, one fixed frame of reference, out of which our universe and any others must have arisen. All of this reality, near and far, is the result of the interactions, the turbulence, the reinforcements and resonances in that field, so other universes could have and probably did occur at many times and places and are occurring still. Some may have been stillborn, some dying in infancy, and given the available time, some may have matured, even died of old age. There is clearly no need for existence of other worlds, other universes in some never-never land. The simple universe permits them to exist right next door, so to speak, but even "next door is too far for us to reach.

Contrary to *the standard models*, I reject the assumptions at their core, those of the homogeneity and isotropy of the universe. These are convenient for the mathematicians, and fuel the speculations, but they are obviously not true, not real. There are regions in the universe that are of higher and lower density, there are gaps in the field, great spaces with apparently only background radiation. To repeat an earlier assertion, reality is rough, not smooth. Only in mathematics, and even there, only in the Platonic ideal, is it smooth.

2. *Does nature have laws?*

The second subject I think I have given short schrift to in this book is the subject of "the laws of nature." (Some of the speculative cosmologists have imagined parallel universes where different "laws" may apply). In one way this notion is part of my concern with the broader subject of the language of physics. I have previously railed about the total tendency to identify the inventions of physicists, the insights, the equations, the constants and limits, as "discoveries," as if they were like the island of Hispaniola, rising out of the sea before Columbus' eyes. Like the word discoveries, the term laws gives in some way the impression that these somehow have always existed and by some mysterious power have determined how the universe has arisen and grown to what it is today. In his wonderful book on the metaphysics of existence, *Why Does the World Exist?* Jim Holt, previously cited, discusses this notion thusly:

p.160
(What are the laws of physics? cs)
". . . . where do the laws themselves live? Do they hover over the world like the mind of God, commanding (everything) to exist? Or do they inhere within the world, amounting to a mere summary of what goes on inside it? . . . consider the other possibility—that the laws of physics have no ontological status of their own. On this view, these laws do not hover over the world or exist prior to it in any way. Rather they are merely the most general possible summary of patterns of events within the world. On this view, the planets don't orbit the sun because they "obey" the law of gravity; instead, the law of gravity (or rather, the general theory of rela-

tivity, which superseded it) summarizes a regular pattern in nature, a pattern that includes the planetary orbits.

Suppose the laws of physics—even the deepest laws, those that will make up the hoped-for final theory—are indeed just summaries of what goes on in the world. Then how can those laws explain anything? Perhaps they can't. That was what Ludwig Wittgenstein thought. "the whole modern conception of the world," Wittgenstein wrote in his *Tractatus*, "is founded on the illusion that the so-called laws of nature are the explanations of natural phenomena. Thus people today stop at the laws of nature, treating them as something inviolable, just as God and Fate were treated in past ages."" (2)

As scientists this is our task, to suss out those patterns and try our best to determine whether they apply broadly across disciplines and whether simpler explanations can be found to describe what we have seen. Instead, we have moved with our wrong assumptions and our misattribution of observations into a mess of greater and greater complexity and contradiction, and have compensated for this with the multiplication of speculative, clearly non-scientific theories and models and then building new fictions on those outcomes.

3. Is there a quantum world?

Yet a third subject on which I never seem to be able to say enough is the unbelievable reliance on quantum mechanics among modern physicists. When you ask a physicist this question, you are nearly always confronted with the response that "it is the most successful theory ever developed," in spite of its logical inconsistency and its total absence of clear understandable roots. Even Max Planck was not convinced, although he is generally credited with its conception.

Quantum theories derive their basic name from Planck's "quantum of energy." He found that the mathematics he was applying to his studies of black body radiation didn't seem to work if he assumed that it was a continuous (analog?) phenomenon, so he devised a way to simplify it by using "units" of measure he called quanta. To his credit, he chose a very tiny unit to which he gave the label "h." His subsequent error was, perhaps in not calling out his contemporaries and followers for turning a conjectural unit of measure into a real physical entity, an *object*, so to speak, that no one had any hope of ever identifying in reality. Newton had run into a similar problem in his attempts to find ways to describe and measure motion, which also tended to trace smooth curves and transitions. In his invention of the differential calculus, he assumed that a smooth path could be seen to be made up of an infinite series of individual points and that a mathematical set of rules could then predict what might occur at any given point on that curve. He didn't, however , give those units their own name, nor did he assume that each point had a separate physical existence. They were imaginary points. For Planck, however, the calculus could not seem to accurately describe the apparently smooth transitions of energy from cold to hot and beyond, so a new unit had to be created, which led to an enormous amount of collateral damage to the intellectual foundations of theoretical physics. I'll explain.

Part of the blame must fall on Isaac Newton, not for any intentional deceit or error, but actually for his brilliance in formulating the laws of force and motion. Long a mystery, with few adherents to any clear and understandable models, Newton's solutions were so self-evident and useful that his picture of a mechanical world became the long term mental model we are stuck in today, one of a world made up of smaller and smaller parts, (atoms and then particles), interacting according to mechanical rules and ending up in mechanically describable patterns. The pervasiveness of a mechanical mindset was, of course not started by Newton. It had been around forever, or at least as long as anyone could remember or was recorded.

So, while Planck's quanta helped the physicists of the microcosmos to a new way of looking at the tiny scale of that world, their way of modeling it was still one of simple mechanics. I have long thought that if they had called their 1920's theory *the mechanics of quanta*, instead of quantum mechanics, this anomaly would have stood out more clearly, and we might have seen it as it really was, a Newtonian description of the micro world. And this is what it is, tiny objects interacting with one another, with forces pushing them together or apart, spinning them around each other or flinging them in all directions, impacting, exploding, leaving nothing behind but their tracks. The "forces" driving these activities are not, of course, those that Newton knew, of laborers pushing, donkeys pulling, gunpowder propelling, taut bowstrings flinging. No, the forces driving the mechanics of quanta were more ephemeral, more difficult to see or feel, and in that they were more akin to those mystical phenomena of eastern, and some western, philosophies and religions, which may have opened the doors to their emergence as some of what we see as the contradictions and paradoxes of quantum theories in general. By the time of the emergence of quantum physics we had already seen the discovery of electromagnetic fields and their attendant forces, but even the leading lights of that revolution, people like Faraday, Maxwell, and Lorentz, in their detailed descriptions of those forces fell back on mechanical analogies, particles, vortices, and the like.

Visualizations of electromagnetic fields even now take the form of two-dimensional sine waves, like these, or three-

293

dimensional sine waves (that is, at right angles to each other), sine waves with spherical polarizations, elliptical polarizations, or compression waves, as in sound waves in the atmosphere. All is mechanical.

So, the world of quantum mechanics is a world of mechanical behaviors, except where those behaviors cannot explain a phenomenon, as in the contradiction between the theory of light as particle, photons, and its apparent behavior as a wave-like entity, in the double slit experiment. Light as a particle was not a concept that could be abandoned, so a mystical conversion, right out of the mystics' textbooks, was the answer. Thus was born the concept of duality, particles that sometimes appear as hard little solid objects and sometimes as wave-like phenomena, a thing called a *wavefunction*. This is, of course, another thing that has never been seen, but it must have happened because how else could this change, from the supposed particle into the wave-like event have happened? One is surprised that its symbol is not the yin and yang of the mystics. A question: when this object is a particle, do the classical rules of mechanics govern its behavior? And when it is a wave, does a whole new set come into play?

A principal mechanical attribute of a particle is its mass, usually denoted a Mev or Gev, meaning thousands or millions of electron volts. But wait, aren't those measures of energy? Of course, but remember $E = mc^2$, so energy equals mass. Then of course, these massive particles are always in motion, so they thus acquire another classical property, momentum, because Momentum = mass x velocity. But there are exceptions. It seems that photons, the particles that make up light, are massless, so how can they have momentum? We know they do because they impart energy to anything they strike. They travel at the velocity of light and pass energy on when blocked, so please explain this, Mr. Physicist. Some of these particles carry something called a charge. Plus charges and minus charges. What is this? Well, it seems that a charge is a property of an entity that causes electrical phenomena. And electrical phenomena are the things that are caused by charged entities. Simple! Now anyone can understand it. Don't even ask what spin is.

Quantum theories are a vain attempt to impose order on a part of the world that resists, even rejects order. There probably actually exist substantive identifiable primal elements of which the universe is made up, but as soon as more that two of these primal elements combine or resonate, the number of possible configurations increases exponentially. If we persist in describing as "fundamental" the entities we are fond of calling "particles," there is no limit to the number of different ones that will be discovered or hypothesized. Accepting this extent of the potentially almost unlimited variety of outcomes is the first step to understanding the physics of *the simple universe*.

Hegel suggests in his metaphysical speculations that being and nothingness are identical and in effect cancel each other, He puts his eggs therefore into the basket of "becoming" as the basis for all existence. The electromagnetic ether of *the simple universe* is the physical equivalent of Hegel's metaphysical "becoming." It is simultaneously the potential and the real source of all that exists.

* * * * * * * *

Does anyone else think this way about the universe, about the microcosmos, the ZMD, the macrocosmos? Well, yes, in parts, some of them truly fractional. Some are, unfortunately still in the crackpot class, with mystical overtones and mechanistic details. Many have seen partway into a new model, but when you read them closely, they still espouse particles as the bottom rung of their theories or say all is magnetism, even without explaining what magnetism is or how it arises. The language limits us. Newton's *forces* have become quantum mechanics' *interactions*. Spin means one thing in the ZMD but an unrecognizable other in QM. Inertia and momentum mean different things. And as we have mentioned previously, laws, constants, limits, and concepts like infinity mean different things as you look at them mathematically or in logical constructions. There have been several recent attempts to point out the inconsistencies and deficiencies of modern theories. Jim Baggott's *Back to Reality*; Alexander Unzicker's *Bankrupting Physics*; Peter Woit's *Not Even Wrong* (mostly about string theory); Even in part John Moffat's *Reinventing Gravity*; but none have postulated an even partly complete alternative reading of the facts and observations. In small part, Unzicker has touched on the subject. In an earlier reference, he is quoted as considering he possibility that "particles" might be just disturbances in the ether," and in the conclusion to his paper *What can Physics Learn from Continuum Mechanics? he says:*

"The main purpose of this paper was to attack two popular preconceptions among today's physicists.

The first one regards the compatibility of aether theories with the experimental facts of special relativity. It has been given in evidence that not the concept of the aether as such is wrong, but the idea of particles consisting of external material passing through the aether. Rather the aether is a concept that yields special relativity in a quite natural way, provided that topological defects are seen as particles.

Independently from this, topological defects appear interesting, because they have been shown to behave as quantum mechanical particles under various aspects." (3)

In the context of *the simple universe*, one should, of course, look at this in its converse, as: "the behavior of quantum mechanical particles can be seen as that of topological defects (in an inelastic continuum, the ether)."

The conclusion I have come to is that the universe *is* simple; in its origin, its organization, and its rules; but it is is incredibly rich in its fecundity and its variety. As Charles Darwin said at the end of *The Origin of Species*, "***from so simple a beginning endless forms most beautiful and most wonderful have been, and are being, evolved.***"

Finally, what I have tried to accomplish in this volume is to suggest an alternative way of thought, a new way of looking at the facts, basing it insofar as possible on the experiments and observations of the real world, and in the process, to attempt to bring us back from the brink of despair over never knowing the truth.

I have today just finished rereading, correcting errors, recasting thoughts, reimaging outcomes of all of this work, from start to finish. In that process, I have realized that what was begun as a work of science has become a work of philosophy, meaning that its intent has become that of changing how we see the world, not just how it works. That is what it has done for me. I have opened my own eyes. I hope it has done the same for yours.

*There is grandeur in this view of life, with its several powers, having been originally breathed into a few forms or into one; and that, whilst this planet has gone cycling on according to the fixed law of gravity, **from so simple a beginning endless forms most beautiful and most wonderful have been, and are being, evolved.**"*

— Charles Darwin, *The Origin of Species*

Notes and References

References to all quotations in the text are noted here as to their original sources where specifically identifiable, including all such references to quotations from Wikipedia articles. Quotations directly from unattributed portions of Wikipedia articles are identified by the title of the article and its year of access. Current Wikipedia articles (i. e. after the date of publication of this book) may differ in some particulars since Wikipedia is a continuously changeable source.

Understanding the nature and content of notes and references for this book requires a brief explanation. Where published books or book, series, or publications of papers or articles in professional journals or magazines are cited, standard rules for such citations are followed. For footnoted items such as historical figures, theories, ideas or concepts, the note/reference will generally contain a brief biographical note or a summary, often drawn from a source such as Wikipedia, often with the specific Wikipedia or other link included for those interested in following up in more detail. In general, the idea has been to avoid filling the narrative pages with great detail about people and things thought to be part of general knowledge, but to provide a reputable source for that detail should the reader desire it.

Notes and references are organized by section and by chapter of the book. If a publication has been previously cited, any new reference will refer back, as completely as is reasonable, with author, section, and chapter as previously referenced. In some instances, notes and references may suggest additional resources, other recommended readings, and supporting authors, whose support may be seen as in line with the ideas expressed in this book but who might not have followed those ideas to the same conclusions reached here.

A note about Wikipedia. The author has been impressed by the completeness and accuracy of most of the Wikipedia entries he has sought out, referenced, and depended upon for biographical and historical notes. Some are more complete than others, but some may turn out to be less reliable. They are included here because they form a quick reference, easy to access and can be keys to further research for any reader so inclined. We have made a point of checking those used against other, corroborating sources and believe them to be quite accurate.

Preface

1   Janna Levin, *How the Universe Got Its Spots*, Anchor Books, New York, 2002

2   Thomas Kuhn, *The Structure of Scientific Revolutions*, The University of Chicago Press, Chicago, 1962, 1970

3   *Standard Model*, Wikipedia, the free encyclopedia, 2013

4   Wikipedia, op. cit.

5   *Big Bang*, Wikipedia, the free encyclopedia, 2014

6   Wikipedia, op. cit.

7   Monseigneur Georges Henri Joseph Édouard Lemaître, (French: 17 July 1894 – 20 June 1966) was a Belgian priest, astronomer and professor of physics at the French section of the Catholic University of Leuven. He was the first known academic to propose the theory of the expansion of the universe, widely misattributed to Edwin Hubble.[2][3] He was also the first to derive what is now known as Hubble's law and made the first estimation of what is now called the Hubble constant, which he published in 1927, two years before Hubble's article.[Lemaître also proposed what became known as the Big Bang theory of the origin of the Universe, which he called his "hypothesis of the primeval atom" or the "Cosmic Egg". ("Obituary: Georges Lemaitre". *Physics Today* **19** (9): 119. September 1966.) (W)

8   Van Flandern, Thomas, Ph.D, "The Top 30 Problems with the Big Bang," www.metaresearch.org

9   Janna Levin, *A Madman Dreams of Turing Machines*, Alfred A. Knopf, New York, 2006

Part 1. The Parts—the Universe as an Ill-structured Problem

1.1 Introduction

1   Robert Boyle, *The Works of the Honourable Robert Boyle*, ed. Thomas Birch, 2nd edn., 6 vols. (London, 1772), III, 316; quoted

in E.A. Burtt, *The Metaphysical Foundations of Modern Science* (Garden City, New York: Doubleday & Company, 1954), 191-192. (W)

2   Newton, Isaac: *Opticks* (1704). Fourth edition of 1730. (Republished 1952 (Dover: New York), with commentary by Bernard Cohen, Albert Einstein, and Edmund Whittaker). (W)

3   Lorentz, Hendrik Antoon (1892), *De relatieve beweging van de aarde en den aether* [*The Relative Motion of the Earth and the Aether*], *Zittingsverlag Akad. V. Wet. 1*: 74–79 (W)

4   *Michelson—Morley experiment*, Wikipedia, the free encyclopedia, 2014

5   Hendrik Lorentz, "The Theory of Electrons and the Propagation of Light," Nobel Prize acceptance address, 1902

6   Max Planck, Wikipedia, the free encyclopedia, 2014. For a solid approach to the complexity of Planck's intellectual motivations for the quantum, for his reluctant acceptance of its implications, see Helge Kragh, Max Planck: the reluctant revolutionary, *Physics World*. December 2000.

7   A. Einstein and L. Infeld, *The Evolution of Physics*, Schuster, NY, (1961)

8   A. Einstein, *Letter to H. A. Lorentz*, 15 November 1919,

9   A. Einstein, *Ether and Relativity Theory*, 5 May 1920, an der Reichs-Universität zu Leiden, Springer, Berlin (1920)

10   A. Einstein, *Forum Philosophicum* 1, 180 (1930)

11   A Einstein, op. cit.

12   F. Selleri, Dipartimento di Fisica, Universita di Bari, "Relativistic physics from paradoxes to good sense -1", in *Ether, Space-time, and Cosmology*, Volume 2, 201-266, Duffy and Levy, Eds., Montreal (2009)

13   Electromagnetic field, Wikipedia, the free encyclopedia, 2014

14   Wikipedia, ibid.

15   Wikipedia, ibid.

1.3 *Disappearing the Particle*

1   Lucretius, *On the Nature of Things*, 1, 157-184, Loeb Classical Library, Harvard (1924)

2   Lucretius, ibid, 213-239

3   Lucretius, ibid. 329-347
4   Lucretius, ibid. 524-548
5   Robert Boyle, *The Works of the Honourable Robert Boyle*, ed. Thomas Birch, 2nd edn., 6 vols. (London, 1772), III, 316; quoted in E.A. Burtt, *The Metaphysical Foundations of Modern Science* (Garden City, New York: Doubleday & Company, 1954), 191-192. (W)
6   *Antoine Lavoisier, 1742-1794,* Wikipedia, the free encyclopedia, 2014
7   *John Dalton* FRS (6 September 1766 – 27 July 1844) was an English chemist, meteorologist and physicist. Wikipedia, the free encyclopedia, 2014
8   *Johann Josef Loschmidt* (15 March 1821 – 8 July 1895), Wikipedia
9   (Brownian motion) *van der Pas, Peter W.* (1971). "The Discovery of Brownian motion". *Scientiarum Historia 13*: 17. (W)
10 *Sir Joseph John "J. J." Thomson*, OM, FRS[1] (/'tɒmsən/; 18 December 1856 – 30 August 1940), Wikipedia
11 *Bohr, Niels* (1913). "On the Constitution of Atoms and Molecules, Part III Systems containing several nuclei". *Philosophical Magazine 26* (155): 857–875. (W)
12 *Atom*, Wikipedia, the free encyclopedia, 2014
13 *Atom*, ibid. (W)
14 John W. Moffat, *Cracking the Particle Code of the Universe*, Oxford University Press, New york, 2014
15 Young, T. (1807). *A Course of Lectures on Natural Philosophy and the Mechanical Arts*. Vol. 1. William Savage. Lecture 39, pp. 463–464. (W)
16 Juan Miguel Marin, *Mysticism in Quantum Mechanics: the forgotten controversy*, European Journal of Physics, *30* (2009) 807–822, May 2009
17 Wave-particle Duality, lgsims96, *HubPages*, worldwide web, (March 2013)
18 Michael Faraday, *Effects of Magnetism on Light*, 1845
19 Hendrik Lorentz, Nobel Prize acceptance lecture, 1902
20 A. Einstein, *On the Electrodynamics of Moving Bodies*, (1905)
21 A. Einstein, *Ether and Relativity*, Lecture at Leiden University, May 5, 1920
22 Alexander Unzicker, *Bankrupting Physics*, Springer-Verlag, Heidelberg, Pangrave Macmillan, New York (2013)
23 James Gleick, *Chaos*, Viking-Penguin, New York, 1987, p. 49
24 Amir Alexander, *Scientific American*, (April 2014)

1.4 The Alternative Models

1   Elizabeth Rosner, *The Speed of Light*, a novel, Ballantine, New York, 2001
2   *Proton*, Wikipedia, the free encyclopedia, 2014
3   http://rcasau.wordpress.com/physics-classification/particle-physics/neutron/
4   *Non-standard cosmology*, Wikipedia, the free encyclopedia, 2014
5   Wikipedia, ibid.
6   file://localhost/(http/::www.allaboutscience.org:big-bang-theory.htm)
7   *Non-standard cosmology*, op. cit.
8   *Non-standard cosmology*, op. cit.

Part 2. The Assemblage/The Universe as a Self-organized System

2.1 How Did It Start?

1   James Gleick, *Chaos*, op. cit.
2   Ludwig Von Bertanffy, *General System Theory*, Braziller, New York, 1969
3   *Chaos Theory*, Wikipedia, he free encyclopedia, 2014
4   Benoit Mandelbrot, *The Fractal Geometry of Nature*, W. H. Freeman & Company, New York, 1977
5   Elmenreich and Fehervari, *Evolving Self-Organizing Cellular Automata based on Neural Network Genotypes*, Springer Science+Business Media, 2011
6   *Self-Organization*, Wikipedia, the free encyclopedia, 2014
7   *Self-organization*, Hermann Haken, Institute for Theoretical Physics I, Center of Synergetics, University of Stuttgart, Germany, Scholarpedia.org, 2014
8   Haken, ibid.
9   Erich Jantsch, *The Self-organizing Universe*, Pergamon Press, 1980
10   *Chaos Theory*, Wikipedia, the free encyclopedia, 2014

11 Bak, P., Tang, C. and Wiesenfeld, K. (1987). "Self-organized criticality: an explanation of 1/f noise". *Physical Review Letters 59* (4): 381–384, 1987

12 Bak, ibid.

13 Per Bak, *How Nature Works*, Springer Science+Business Media, New York, 1996

14 *Degrees of freedom* (Physics and Chemistry), Wikipedia, the free encyclopedia, 2014

15 Allen B. Downey, *Think Complexity: Complexity Science and Computational Modeling*, O'reilly Media, Inc., Sebastopol, CA, 2012

16 Markus J. Aschwanden, Self-criticality Systems in Astrophysics, http://arxiv.org/abs/1207.4413

17 *Resonance*, Wikipedia, the free encyclopedia, 2014

2.2   Cosmic Evolution

1   Andrew Knoll, *Interview on NOVA, July 1, 2004*

2   Hilton Ratcliffe, *The Static Universe, Exploding the myth of Cosmic Expansion,,* Apeiron, Montyreal, 2010

3   Ratcliffe, ibid.

4   *Chaos Theory*, Wikipedia, the free encyclopedia, 2014

5   *Fractal*, Wikipedia, the free encyclopedia, 2014

6   Wikipedia, ibid.

7   Stephen Wolfram, *A New Kind of Science*, Wolfram Media, Inc., Champaign, IL, 1959

8   Andrew Ilachinski, *Cellular Automata*, World Scientific Publishing, Singapore, 2001

9   Gabriele Rossi, *The Mathematics of the Models of Reference*, 2010

10 *Cellular Automaton*, Wikipedia, the free encyclopedia, 2014

11 *Phase Transition*, Wikipedia, the free encyclopedia, 2014

12 *Kenneth Wilson*, Wikipedia, the free encyclopedia, 2014

2.3 Summary: The Simple Universe

Part 3. The User Manual/FAQ

3.1 Introduction

1   Gregory Bateson, *Steps to an Ecology of Mind*, BallantineBooks, New York, 1972

3.2 Microcosmos/Imagining the Unseeable

1   *Electromagnetic spectrum*, Wikipedia, the free encyclopedia, 2014
2   *Visible spectrum*, Wikipedia, the free encyclopedia, 2014
3   Young, Thomas (1804). "Bakerian Lecture: Experiments and calculations relative to physical optics". *Philosophical Transactions of the Royal Society* **94**: 1–16. (*Thomas Young (scientist)*, Wikipedia, the free encyclopedia, 2014)
4   *Planck constant*, Wikipedia, the free encyclopedia, 2014
5   Andrew Zimmerman Jones, "Photoelectric Effect Explained," physics.about.com (2014)
6   Francesco Maria Grimaldi, *Physico mathesis de lumine, coloribus, et iride, aliisque annexis libri duo* (Bologna ("Bonomia"), Italy: Vittorio Bonati, 1665) from *Diffraction*, Wikipedia, the free encyclopedia, 2014
7   *Isaac Newton*, by Wikipedians, Google Books, Pedal Press, 2014, p.8
8   *Refractive index*, Wikipedia, the free encyclopedia, 2014
9   Daniel J. Castellano, "Ontological Interpretation of Quantum Mechanics," (2011)
     http://www.arcaneknowledge.org/science/quantum.htm#top

3.3 ZMD/ the zone of middle dimensions

1   Jeff Hawkins, *On Intelligence,* Henry Holt and Company, LLC, New York, 2004
2   *Robert Boyle*, Wikipedia, the free encyclopedia, 2014

3   Newlands, John A. R. (20 August 1864). "On Relations Among the Equivalents". *Chemical News* **10**: 94–95.
     Newlands, John A. R. (18 August 1865). "On the Law of Octaves". *Chemical News* **12**: 83.
     (*Periodic table*, Wikipedia, the free encyclopedia, 2014)
4   Walter Murch, *In the Blink of an Eye,* Silman-James Publishers, 2001
5   *"In all cases in which work is produced by the agency of heat, a quantity of heat is consumed which is proportional to the work done; and conversely, by the expenditure of an equal quantity of work an equal quantity of heat is produced.* Clausius, R. (1850), page 373, translation here taken from Truesdell, C. A. (1980), pp. 188–189. From: First Law of Thermodynamics, Wikipedia, the free encyclopedia, 2014
6   A. Einstein, "On the Electromagnetics of Moving Bodies," *Annelen der Physik*, 1905

3.4 macrocosmos/imagining the unreachable

1   Chris Impey, *How it Began*, Norton, New York, 2012
2   *Star Formation*, Wikipedia, the free encyclopedia, 2014
3   "Small 'helper' stars needed for massive star formation," *UCBerkeleyNews,*
     http://berkeley.edu/news/media/releases/2008/02/27_helperstar.shtml
4   *Millenium Soimulation Project*, Max Planck Institute for Astrophysik, http://www.mpa-garching.mpg.de/galform/virgo/millennium, 2014
5   Zhuravleva , et al, *Turbulent Heating in Galaxy Clusters Brightest in X-rays*, arXiv.1410.6485v1
6   J. W. Moffat, "Gravitational Theory, Galaxy Rotation Curves and Cosmology without Dark Matter," *JCAP* 0505 (2005) 003, arXiv: astro-ph/0412195
7   A. Einstein, "Ether and the Theory of Relativity," Address at the University of Leiden, May 5, 1920
8   M. Lockwood, R. Stamper, and M.N. Wild, "A *Doubling of the Sun's Coronal Magnetic Field during the Last 100 Years,*" *NATURE* (Vol. 399, 3 June 1999. Pages 437-439),
9   John W. Moffat, *Reinventing Gravity*,   Collins/Smithsonian, New York, 2008, p. 33
10   Moffat, ibid. p. 201

11   E.P. Verlinde. "On the Origin of Gravity and the Laws of Newton".
      JHEP. arXiv:1001.0785, 2011

12   *Magnetism*, Wikipedia, the free encyclopedia, 2014

13   *Magnetic Field*, Wikipedia, the free encyclopedia, 2014

14   Annie Sneed, "Earth's Impending Magnetic Flip," *Scientific American,*
      September 16, 2014

3.5. misattribution/how we got it wrong

1   Jim Baggott, *Farewell to Reality*, Pegasus Books, New York, 2013

2   *Quark*, Wikipedia, the free encyclopedia, 2014

3   Alfred Korzybski, *Science and Sanity: An Introduction to Non-
      Aristotelian Systems and General Semantics,* Institute of General Se-
      mantics; 5th edition (April 1, 1995)

4   Bertrand Russell, Arthur North Whitehead, *Principia Mathematica,*
      Second Edition, 1927

5   Juan Miguel Marin, "Mysticism" in Quantum Mechanics: The For-
      gotten Controversy," *European Journal of Physics*, v30 n4 p807-822
      Jul 2009

6   Benoit B. Mandelbrot, *Fractals and Chaos: The Mandelbrot Set and
      Beyond*, Springer, 2004

7   J. M. McDonough, "Introductory Lectures on Turbulence, Physics,
      Mathematics, and Modeling," University of Kentucky,
      http://www.engr.uky.edu/~acfd/lctr-notes634.pdf

8   Gregory L. Eyink, "Stochastic flux freezing and magnetic dynamo"
      Phys. Rev. E 83, 056405 – Published 27 May 2011

9   Vlad Tarko, "Turbulence - the last mystery of classical physics,
       http://archive.news.softpedia.com/news/Turbulence-the-last-mystery-
      of-classical-physics-17327.shtml, 11Feb 2006

Part 4. Afterword

1   Jim Holt, *Why Does the World Exist?*, Liveright Publishing, New
      York, 2012

2   Jim Holt, ibid.

3   Alexander Unzicker, "What can Physics learn from Continuum Me-
      chanics ?" arXiv-gr-qc/0011064

Index

imagine darkness

www.ingramcontent.com/pod-product-compliance
Lightning Source LLC
Chambersburg PA
CBHW051852170526
45168CB00001B/82